THE SPITFIRE
AN ICON OF THE
SKIES

PHILIP KAPLAN

Pen & Sword
AVIATION

First printed in Great Britain in 2017 by
Pen & Sword Aviation

An imprint of Pen & Sword Books Ltd.
47 Church Street
Barnsley
South Yorkshire
S70 2AS

A CIP record for this book is available from the British Library.

ISBN 978 1 47389 852 3

Printed and bound in India
by Replika Press, Pvt. Ltd.

Pen & Sword Books incorporates the Imprints of Pen & Sword Archeology, Atlas, Aviation,
Battleground, Discovery, Family History, History, Maritime, Military, Naval, Politics, Railways,
Select, Transport, True Crime, Fiction, Frontline Books, Leo Cooper, Praetorian Press.
Seaforth Publishing, Wharncliffe and White Owl.

For a complete list of Pen & Sword titles please contact:
PEN & SWORD BOOKS LIMITED
47 Church Street
Barnsley
South Yorkshire
S70 2AS
England

E-mail: enquiries@pen-and-sword.co.uk
Website: www.pen-and.sword.co.uk

Book design: Philip Kaplan

THE MK XVI SPITFIRE OF KERMIT
WEEKS WAS RESTORED BY TONY
BIANCHI AND HIS COMPANY,
PERSONAL PLANE SERVICES AT
BOOKER AIRFIELD, ENGLAND.

PS890, A FORMER ROYAL
THAI AIR FORCE PRXIX.

CONTENTS

PS853, A PHOTO-RECONNAISSANCE PRXIX SPITFIRE BUILT IN 1945 AT SOUTHAMPTON. IT WAS PURCHASED BY ROLLS-ROYCE IN 1996.

"When you look at a Spitfire you see a nation at its finest" wrote the American Editor-in-Chief of *Flight Journal* Budd Davisson. "When I look at a Spitfire I see art deco perfection with wings; a mechanical object shaped with fewer compromises than most things made by man. I see something drenched in history, a chariot for The Few. The Spitfire is a work of art, whether or not it was intended to be. The one thing it is not is just another aeroplane."

Another of its admirers once wrote: ". . . the Spitfire is as much a British national hero as Wellington, Nelson, or Montgomery. It has become the most recognisable icon of the Second World War for several generations of Britons. From the throaty growl of its Rolls-Royce Merlin or Griffin to its beautifully-tapered elliptical wings, the Spitfire is a true aeronautical thoroughbred. Regarded by many as the saviour of the nation in its darkest hour, the Spitfire is without doubt the most famous combat aircraft ever produced in Britain and probably anywhere else."

In June 1936, Supermarine Aviation Works, Southampton, England, was awarded an order from the British Air Ministry for the production of 310 Spitfire aircraft. It was, by a large margin, the biggest production order the company had ever received. By the end of production in all marks, 22,759 Spitfires and Seafires had been built. The Spitfire was the only Allied fighter that remained in full production from the very first day of the Second World War to the very last day, and it was retained in front-line service with the Royal Air Force until 1951.

An example of the early Spitfires remains airworthy today, the Mk 1a AR213, which was built by Westland at Yeovil, Somerset, in late 1940 and delivered to the Royal Air Force in February 1941. It survives today largely because it was never exposed to the hazards of air combat—only to those of Nos 53 and 57 Operational Training Units. At 57 OTU it was, for a time, the personal aircraft of James 'Ginger' Lacey during his stint there as an instructor. Lacey had been credited with downing more enemy aircraft in the Battle of Britain than any other RAF pilot.

AR213 has passed through a succession of owners since the war years, and has been cared for mostly by Doug and Tony Bianchi of Personal Plane Services at Wycombe Air Park, Buckinghamshire. Tony maintained and flew the Spitfire since the 1970s and has come to know the aeroplane, its strengths and weaknesses, characteristics and quirks, better than anyone in its long history. They have shared many adventures and a few hair-raising moments, and he has long appreciated what may be its most outstanding characteristic—as one of the lightest Spitfires ever built. With a superb power-to-weight ratio, it handles—in his informed opinion—better than most (if not all) subsequent Spitfire marks. Considering the Spitfire's reputation for wonderful handling qualities, that is saying a lot.

While flown and well-maintained for most of its long career, AR213 had never undergone a total restoration until relatively recently. In 2002 it was decided to have the aeroplane taken right down to the last rivet and restored to the highest possible standard. It emerged from the project a beautiful, pristine representation of a factory-fresh 1941 Mk 1, looking, feeling, smelling, and handling exactly as Tony knew it should.

Progress during the restoration was at times agonisingly slow, and the ample skills of PPS team members Tom Woodhouse, Franco Tambascia, Tony Bird and others were frequently put on hold while they awaited the arrival of custom-made or hard-to-find structures and equipment sourced from suppliers the world over. Nothing was spared and no corners were cut in the effort to achieve the desired result and, finally, in the spring of 2008, the job was finished.

The Spitfire An Icon of the Skies looks at both the magnificent restoration of AR213, and at the Spitfire generically. It considers the mystique and charisma associated with the type, its principal designer R.J. Mitchell, the Spitfires of the pre-war years, the Spitfire in the Battle of Britain, flying the aeroplane, the roles of the Spitfire in the Second World War, the amazing career of Alex Henshaw as Chief Test Pilot for the production programme at Castle Bromwich, Birmingham, the famous Rolls-Royce Merlin engine that powered so many Spitfires, some of the motion picture and television performances of the Spitfire, and the phenomenal evolution of the warbird movement.

When the German Chancellor Adolf Hitler ordered preparations for his Operation Sea Lion—the invasion of England, his forces were fresh from their triumphant Blitzkrieg, or Lightning War, on the European Continent in 1940. The foundation for such an invasion required the destruction of British coastal and air installations and aircraft factories by bombing. The odds seemed to favour the German Air Force after the fall of France to Hitler's army. The Luftwaffe, his vaunted air arm, had both a significant numerical advantage over the Royal Air Force in aircraft and aircrew, and a number of confident and seasoned veterans of the Spanish Civil War, Poland, and the Western Front campaigns. In the Messerschmitt Bf 109, the Germans had what was then considered the best fighter in the world, soon to face the untested, unproven Hawker Hurricane and the Supermarine Spitfire of the RAF, which would have to field a far smaller force of eager and highly-motivated but mostly under-trained pilots.

What the British did have in their arsenal though was the advanced electronic weapon—radar, which gave them early warning of enemy bombing attacks, enabling them to get fighter defences into the air to quickly intercept the German raiders. And they had the marvellous Lord Beaverbrook, the new Prime Minister Winston Churchill's boss of aircraft production, seeing that RAF Fighter Command was never short of replacement aircraft throughout the costly fifteen-week Battle of Britain in the summer and fall of that year.

For the pilots of Fighter Command the Battle began in early July, though many historians have focused on August 13th, which the Germans referred to as *Adler Tag* (Eagle Day) as the official start of the campaign. Through the course of the long battle the Spitfires and Hurricanes went up to meet and engage the bombers and fighters of the Luftwaffe and extracted a punishing toll in killed, wounded and captured German airmen and destroyed enemy aircraft. And by October 12th Sea Lion had been called off; the Luftwaffe having failed to accomplish the goals set for it preparatory to the invasion. Hitler then turned his attention eastward to a new campaign . . . against the Soviets. The well-intended but appeasing British Prime Minister Neville Chamberlain had been succeeded in that office by Winston Churchill in May and in that role Churchill's greatest achievement may simply have been the employment of his spirit, expressed as it was with a growling eloquence that served to sustain the British people when they were truly alone and facing the might of Nazi Germany. He told them: "I have nothing to offer but blood, toil, tears and sweat," but in fact he was offering something they needed most, hope. "Upon the Battle of Britain", he warned, "depends the survival of Christian civilisation . . . Hitler knows that he will have to break us in this island or lose the war . . . Let us therefore brace ourselves . . . that if the British Empire and its Commonwealth last for a thousand years, men will still say 'This was their finest hour.'"

In *The Spitfire Log* Peter Haining wrote: "The Spitfire has, with every reason, earned a special place in aviation history as well as in the hearts of ordinary men and women, not just because it symbolised a nation's hopes for freedom, but because it flew so superbly and came close to embodying the kind of grace and elegance that mankind has so admired in the birds since the dawn of history. In a way, it was our dreams given wings—and even in our present age of supersonic missiles and space travel, there is much to respect in the feat of design and engineering which it represents."

"When the Hurricanes and Spitfires were shooting at Jerries over our village, it sounded just like someone was tearing a piece of rough calico."
—Mass Observation, 1940

"The Spitfire didn't win the Battle of Britain, but the battle would have been lost without the Spitfire, and had the battle been lost then we would have lost the war."
—Alex Henshaw, Chief Test Pilot, Castle Bromwich

MYSTIQUE

INSTRUCTORS AND STUDENTS OF THE UNIVERSITY OF LONDON AIR SQUADRON PREPARING FOR AN AFTERNOON'S FLYING IN THEIR AVRO TUTOR BIPLANES.

ON THE PREVIOUS SPREAD, LIKE AR213, AR501 WAS BUILT BY WESTLAND IN SOMERSET, HER POSTWAR CAREER HAS INCLUDED MANY YEARS OF FLYING AS PART OF THE SHUTTLEWORTH GROUP OF AEROPLANES AT OLD WARDEN IN BEDFORDSHIRE.

Certain icons of Britishness powerfully evoke the popular culture and character of that proud and accomplished society—fish and chips, the cup of tea, the double-decker Routemaster bus, the "mini" (both car and skirt), the bright red phone box, the ubiquitous pub, Big Ben, the Beatles, the Bobbies, hedgerows, London taxis, Tower Bridge, the white cliffs of Dover and, unquestionably, the Spitfire. No aeroplane in history has approached the eminence it has achieved both within and outside the aviation world. None has garnered the level of admiration afforded by those who knew the Spitfire first hand, by its adversaries, and by the general public, and none has earned higher praise from the great majority of pilots who have had the privilege of flying it. No single object has contributed more to the survival of the British nation than the Spitfire. During the Battle of Britain, together with its highly capable partner, the Hawker Hurricane, the elegant little fighter prevented the powerful German Air Force from opening the way for a Nazi invasion of England. Had that invasion succeeded, Britain (and later the United States) would have been denied the use of hundreds of English airfields from which to conduct their decisive strategic bombing campaign against Germany and German-occupied Europe. For the Allies the result would have been catastrophic.

What made the Spitfire the object of profound respect and affection that it was and still is? What gave it that special place in the hearts and minds of so many people, young and old? What was it about the handsome fighter plane that so lifted the common spirit

and put wings under the words of Prime Minister Winston Churchill, whose praise for the pilots of Royal Air Force Fighter Command in the midst of the Battle of Britain ended with the famous phrase "Never in the field of human conflict was so much owed by so many to so few"? There are, of course, the endless clichés . . . "if it looks right it will fly right", "a real pilot's aeroplane", "very advanced for its time", "you didn't fly it; you strapped yourself into it and became part of it", "every pilot's favourite", "it was very light on the controls and almost flew itself", "remembered with affection by all who flew her", "honest and straight forward", "a poem of speed and precision", etc.

One former Spitfire pilot recalled: "You feel like Henry the Fifth and Joan of Arc tied to the stake at the same time. You can move your hands, feet and head the few inches that are required; your Spitfire will do the rest. You are the most powerful, the fastest, the most manoeuvrable fighting man in the world." And as the writer Peter Haining put it: ". . . the Spitfire may just have given its pilots the closest feeling to that of being like a bird. And a ferocious bird of prey, to be sure! No wonder a whole generation of young men found it their highest aspiration and greatest pride to be a Spitfire pilot." The Second World War Free French aviator, Pierre Clostermann, wrote: "For a pilot every plane has its own personality, which always reflects that of its designers and colours the mentality of those who take it into action. The Spitfire, for instance, is typically British. Temperate, a perfect compromise of all the qualities required of a fighter, ideally suited to its task of

THE·SCHNEIDER·CUP·AIR·
SEPTEMBER 1929

defence. An essentially reasonable piece of machinery, conceived by cool, precise brains and built by conscientious hands. The Spitfire left such an imprint on those who flew it that when they changed to other types they found it very hard to get acclimatised."

"I know I fell in love with her the moment I was introduced that summer day in 1938. I was captivated by her sheer beauty; she was slimly built with a beautifully proportioned body and graceful curves just where they should be. In every way to every young man—or, in my case, middle-aged man—she looked the dream of what one sought. Mind you, some of her admirers warned me that she was what mother called 'a fast girl', and advised that no liberties should be taken with her until you got better acquainted. I was warned to approach her gently but once safely embraced in her arms I found myself reaching heights of delight I had never before experienced. This was my introduction to that early Spitfire, the consummation of R.J. Mitchell's design work of sixteen years, supported by a team of colleagues without whom, as he was always at pains to stress, he could never have achieved all that he did."
– The Right Honourable The Lord Balfour of Inchyre, PC, MC, Under Secretary of State for Air, 1938-1944

Her pilots related to the Spitfire in a rare and special way. They bonded with her and loved her for the fresh young thing she was, as enamoured as a teenage boy when he meets his dream girl. However, they quickly learned that she was not one to be taken for granted. She demanded great respect and attention but, handled with care, would deliver a spectacular performance.

Al Deere, the great New Zealand fighter pilot credited with the destruction of twenty-two German aircraft, wrote: "On March 6th, 1939, I flew my first Spitfire, an aircraft very little different from the later Marks with which the RAF finished the war. In many respects, however, it was a nicer aircraft and certainly much lighter than subsequent Marks which became over-powered and over-weighted with armour plating and cannons. The transition from slow biplanes to the faster monoplanes was effected without fuss, and in a matter of weeks we were nearly as competent on Spitfires as we had been on Gladiators.

"Two days after my first flight in a Spitfire, I was sent down to Eastleigh airfield near Southampton to collect a new aircraft from the Supermarine Works where the Spitfire was produced. I can still recall the thrill of that momentous occasion. As I sat in the cockpit ready to take-off, I had a feeling of completeness which was never to leave me in the years of combat ahead. Indeed, in a dogfight I always felt that my Spitfire and I were as one—we certainly grew up together.

"The weather was pretty bad for the trip and I had the greatest difficulty finding Hornchurch. Such was my relief when I did find the airfield that I couldn't resist a quick beat-up, very much frowned on in those days. I noticed as I taxied towards the squadron hangar after landing that there was more than the normal complement of pilots to greet me. I was soon to learn why. Not only had I committed the sin of beating up the airfield, I had neglected to change from fine to coarse pitch after take-off from Eastleigh with the result that there was a great deal more noise than normal on my quite small beat-up and worse still, the brand new Spitfire was smothered in oil thrown out by the over-revving engine. A very red-faced Pilot Officer faced an irate 'Bubbles' Love, my flight commander, who was decent enough to let the matter rest at a good 'ticking-off.' "

Another famous Battle of Britain pilot, Group Captain Hugh "Cocky" Dundas, recalled in his excellent book *Flying Start*: "There is something Wagnerian about facets of the Spitfire story, the more so since it is certainly true that there never was a plane so loved by pilots, combining as it did sensitive yet docile handling characteristics with deadly qualities as a fighting machine. Lovely to look at, delightful to fly, the Spitfire became the pride and joy of thousands of young men from practically every country in what, then, constituted the free world. Americans raved about her and wanted to have her; Poles were seduced by her; the Free French undoubtedly wrote love songs about her. And the Germans were envious of her.

"Little did I know as I taxied in from that first Spitfire flight that I would not taxi in from my last until late in 1949. We went through the war together, with only a year's separation when, in 1942, I temporarily—and not very happily—flirted with the Typhoon.

"In all those years no misfortune which came our way was ever the

fault of the Spitfire. Owing to the loss of my second log book—the first ran up to the end of July 1942—I do not know exactly how many hundreds of hours I spent in a Spitfire's cockpit, over sea, desert and mountains, in storm and sunshine, in conditions of great heat and great cold, by day and by night, on the deadly business of war and in the pursuit of pleasure. I do know that the Spitfire never let me down and that on the occasions when we got into trouble together the fault was invariably mine."

Wing Commander George Unwin, twice winner of the Distinguished Flying Medal in the Second World War: "I wouldn't say that we were terrified at the prospect of flying the Spitfire for the first time, but we were more than a little apprehensive to say the least, since we had been flying the most gentlemanly of machines in the Gloster Gauntlet since 1935—the latter was even easier to handle than the RAF's primary training aircraft of the period, the Avro Tutor. It touched

down at about 50 mph, possessed no flaps, no hood and had no retractable anything—indeed, the only thing that moved was the throttle! We therefore went straight from this thoroughly familiar machine one day, straight on to the seemingly massive brute of a Spitfire the next, without the aid of even a dual-control trainer. However, a lot of our initial fears were soon removed following the display put on over our airfield by Supermarine Test Pilot Jeffrey Quill prior to his delivery of the first Spitfire. This performance would have satisfied any potential pilot of the fighter, and he completed his routine by carrying out a perfect landing just to show us how easy it was. Nevertheless, we went about our transition without even a set of pilot's notes in those early days of 1938. All we possessed was a scrap of paper on which had been scrawled the climbing, diving, cruising and stalling speeds, and that was about it. After the Gauntlet the cockpit seemed to be full of switches and levers, operating

things like the flaps and undercarriage that popped open and closed. Indeed, on the first batch of Mk Is which we received, extension and retraction of the gear was achieved through the pumping of a huge lever sited on the right-hand side of the cockpit—you could always tell a Spitfire pilot undergoing his first solo take-off, for as he pumped on the lever he would push the control column up and down as well, causing the aircraft to porpoise accordingly! Fortunately, powered undercarriage retraction was installed in the next production run of aircraft."

In his autobiography, *Smoke Trails in the Sky*, Squadron Leader Tony Bartley remembered his first time with the Spitfire: "On March 6th 1940, three weeks before my 21st birthday, No 92 Squadron was re-equipped with Spitfires. The bogey men of Blenheims and black nights had dawned into horizons of dawn sunlight and blue skies.

LEFT: ICONS OF BRITISH-NESS; RIGHT AND CENTRE: AL DEERE, FAR RIGHT: GEORGE UNWIN. PAGE 16: IN THE ALLEN AIRWAYS FLYING MUSEUM, EL CAJON, CALIFORNIA.

"One of the unique and most alarming experiences in one's life must surely be to find oneself alone in an aeroplane for the very first time, completely dependent upon oneself to get back to mother earth. In air terminology, this is called 'Soloing'. My second most exciting experience was to fly a Spitfire for the first time. It was like driving a racing car after an Austin . . . riding a racehorse, after a hack. It just didn't seem to want to slow down. When one pulled back on the throttle, it took a long time to take effect on its speed.

"In contrast to the Blenheim, the Spitfire was the perfection of a flying machine designed to combat and destroy its enemy. It had no vices, carried great fire power, and a Rolls-Royce motor which very rarely stopped. An aerodynamic master-piece, and a joy to fly."

"I preferred the Spitfire to other fighters because it had few vices. It was fast, very manoeuvrable and had a high rate of climb."
– Squadron Leader P.G. Jameson

In my research for this and earlier books, I have never found a single Spitfire pilot who was less than exuberant about his experience of the aeroplane. Few failed to lavish it with praise and fewer still found any fault with R.J. Mitchell's fighter. That is not to say that some haven't raised an occasional critical point about the machine. One of the very few negative remarks I have come across about the Spitfire originated from Mitchell himself. Surprisingly, he abhorred the name Spitfire. Two months after the historic first flight of K5054, the prototype Spitfire, he was informed that the Air Ministry, which was exceedingly pleased with her performance, had, on the recommendation of Sir Robert McLean, Chairman of Vickers (Aviation) Ltd, decided to call her Spitfire (possibly after the Katharine Hepburn character in the 1934

movie of that title). Mitchell commented: "Just the sort of bloody silly name they would give it." On that point he would be in the minority, as the pilots of Fighter Command, the Royal Air Force, the British public and, eventually, the whole world would come to appreciate the perfect aptness of the name.

Joseph "Mutt" Summers was Chief Test Pilot at Vickers and the man who first flew the prototype Spitfire, on March 6th, 1936 at Eastleigh Airfield, near Southampton. The principal Spitfire test pilot, and Summers' assistant at Vickers, was Jeffrey Quill, who recalled the day: "When Mutt shut down the engine and everybody crowded round the cockpit, with R.J. foremost, Mutt pulled off his helmet and said firmly, 'I don't want anything touched.' This was destined to become a widely misinterpreted remark. What he meant was that there were no snags which required correction or adjustment before he

flew the aircraft again. The remark has crept into folklore, implying that the aeroplane was perfect in every respect from the moment of its first flight, an obviously absurd and impracticable idea. After the fifteen-minute first flight, the aircraft was still largely untested and unproven, having done one take-off and one landing. Mutt was far too experienced a hand to make any such sweeping statement at that stage in the game.

"However, it was a highly successful and encouraging first flight and Mutt Summers, with his experience of flying a great variety of prototype aircraft, was a highly shrewd judge of an aeroplane. By now I knew him well enough to see that he was obviously elated. Certainly to those of us watching from the ground 'the Fighter' in the air took on a very thoroughbred and elegant appearance, a strong but indefinable characteristic which was to remain with it throughout its long, varied and brilliantly successful life as a fighting aeroplane. Later that afternoon I flew Mutt back to Brooklands in the Falcon and we put the aircraft away and walked across to have a drink in Bob Lambert's well-known and congenial Brooklands Flying Club bar. Mutt was pleased, obviously, to have one more successful first flight tucked under his belt, and I felt excited about this long, sleek and elegant machine which I knew that soon I would fly. A hundred yards from where Mutt and I were leaning against the bar was the hangar in which was standing K5083, the prototype Hurricane, which had made its first flight in the hands of George Bulman some four months previously.

"So the two new fighter aircraft—destined four years later to save our country in time of war—had now both flown in prototype form. Neither was yet anywhere near being a practical fighting machine nor was either yet ordered in quantity by the Royal Air Force, so much work still remained to be done. Ironically perhaps, the very next day, 7th March, Hitler's troops re-entered the demilitarised zone of the Rhineland in direct defiance of the Versailles Treaty. France and Britain, paralysed by political indecision, did nothing. Had they reacted with even the slightest resolution or show of military force the German Army was under orders to withdraw immediately, but that, of course, was not known then. Thus the last chance of effectively and cheaply blocking Hitler's expansionist ambitions was lost and the Spitfire was born into the inevitability of war."

"It was always a delight to fly, was supremely responsive to the controls at all speeds in all attitudes of flight and, with all this, a very stable gun platform."
– Pilot Officer P.H. Hugo

"The versatility and the deceptive toughness of this fighter made it, I think without question, the outstanding fighter aircraft of the Second World War."
– Air Commodore Al Deere

The greatest German fighter leader, a superb fighter pilot and high-achieving aerial hunter of the Luftwaffe, General Adolf Galland admired the Spitfire and was never shy about his opinions. In *The First and the Last*, his perspective of the German air force in the Second

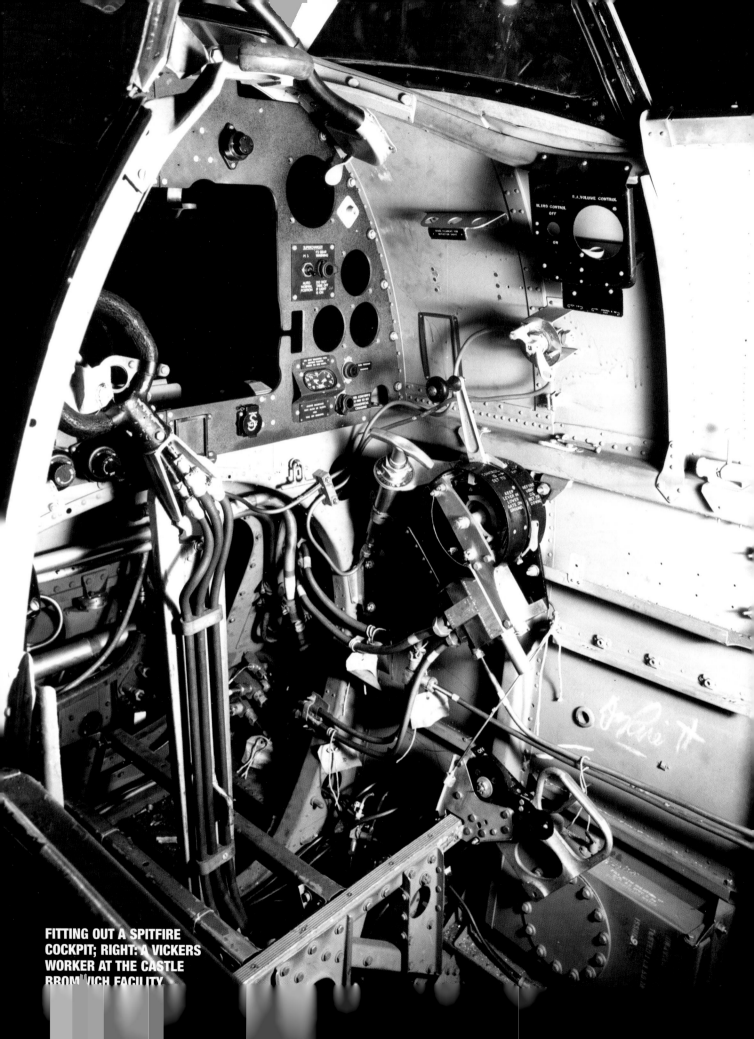

FITTING OUT A SPITFIRE
COCKPIT; RIGHT: A VICKERS
WORKER AT THE CASTLE
BROMWICH FACILITY

World War, he wrote: "The second phase of the Battle of Britain, lasting from July 24th to August 8th, 1940, was essentially a fighter battle. On its opening day I was with my wing for the first time in action over England. Over the Thames Estuary we got involved in a heavy scrap with Spitfires, which were screening a convoy. Together with the Staff Flight, I selected one formation as our prey, and we made a surprise attack from a favourably higher altitude. I glued myself to the tail of the plane flying outside on the left flank and when, during a right-handed turn, I managed to get in a long burst, the Spitfire went down almost vertically. I followed it until the cockpit cover came flying towards me and the pilot baled out, then followed him down until he crashed into the water. His parachute had failed to open.

"The modern Vickers Supermarine Spitfires were slower than our planes by about 10 to 15 mph, but could perform steeper and tighter turns. The older Hawker Hurricane, which was at that time still frequently used by the British, compared badly with our Me 109 as regards speed and rate of climb. Our armament and ammunition were also undoubtedly better. Another advantage was that our engines had injection pumps instead of the carburettors used by the British, and therefore did not conk out through lack of acceleration in critical moments during combat. The British fighters usually tried to shake off pursuit by a half-roll or half-roll on top of a loop, while we simply went straight for them, with wide-open throttle and eyes bulging out of their sockets.

"During this first action we lost two aircraft. That was bad, although at the same time we had three confirmed kills. We were no longer in doubt that the RAF would prove a most formidable opponent.

"It was in August that Reichsmarschall Göring came to visit us on the coast of France. The large-scale attacks of our bombers on Britain were imminent, and the air supremacy necessary for them had not been achieved to the degree expected. The British fighter force was wounded, it was true, but not beaten. And our pursuit Stuka and fighter force had naturally suffered grievous losses in material, personnel and morale. The uncertainty about the continuation of the air

offensive reflected itself down to the last pilot. Göring refused to understand that his Luftwaffe, this shining and so far successful sword, threatened to turn blunt in his hand. He believed there was not enough fighting spirit and a lack of confidence in ultimate victory. By personally taking a hand, he hoped to get the best out of us.

"To my mind he went about it the wrong way. He had nothing but reproaches for the fighter force, and he expressed his dissatisfaction in the harshest terms. The theme of fighter protection was chewed over again and again. Göring clearly represented the point of view of the bombers and demanded close and rigid protection. The bomber, he said, was more important than record bag figures. I tried to point out that the Me 109 was superior in the attack and not so suitable for purely defensive purposes as the Spitfire, which, although a little slower, was much more manoeuvrable. He rejected my objection. We received many more harsh words. Finally, as his time ran short, he grew more amiable and asked what were the requirements for our squadrons. Mölders asked for a series of Me 109s with more powerful engines. The request was granted. 'And you?' Göring turned to me. I did not hesitate long. 'I should like an outfit of Spitfires for my group.' After blurting this out, I had rather a shock, for it was not really meant that way. Of course, fundamentally I preferred our Me 109 to the Spitfire, but I was unbelievably vexed at the lack of understanding and the stubbornness with which the command gave us orders we could not execute—as a result of many shortcomings for

which we were not to blame. Such brazen-faced impudence made even Göring speechless. He stamped off, growling as he went."

Günther Rall, the German World War Two ace and third highest-scoring fighter pilot in history, was flying from a Channel coast base near Calais in the summer of 1940: "We were facing the Spitfires in very rough dogfights every day." He felt hampered by the Luftwaffe command requirement that his Messerschmitt 109 pilots fly as close escort for the German bombers and the slow Ju-87 Stuka dive-bombers. Rall believed that the requirement made the German fighter

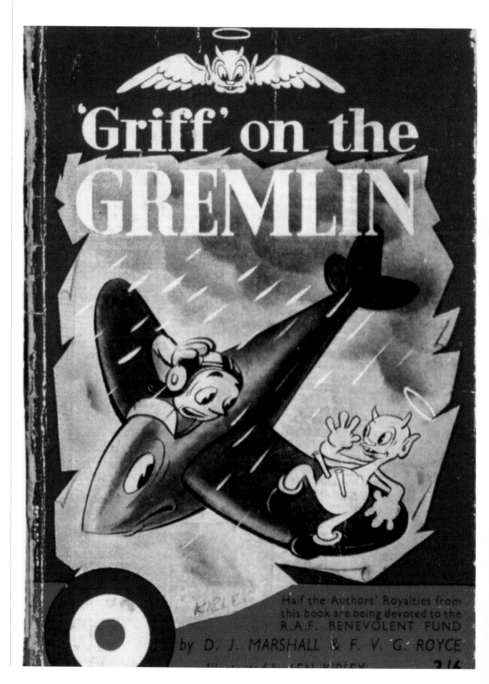

pilots sitting ducks and provided the Spitfires with a clear advantage, making it easier for the RAF pilots to bounce the 109s from a higher altitude and destroy them. The leadership of Rall's outfit, JG-52, suffered considerably in the early days of the Battle of Britain, losing its commanding officer and two squadron leaders during its first four missions to England. These losses, however, led to Rall's promotion to squadron leader.

The brilliant fighter leader and commander of No 92 Squadron Spitfire pilots at Biggin Hill in the Battle of Britain, Group Captain Brian Kingcome remembered in *A Willingness to Die*: "The RAF offered not only a permanent career but a more exciting and rewarding one than I could have believed possible. Actually to be paid to fly the best aeroplanes in the world—it was a notion surpassing my wildest dreams.

"The Battle of Britain had been fought and won three months earlier, and Leslie Howard, the country's best-loved, best-known film actor, was directing and acting in his film *The First of the Few*, based on the life of R.J. Mitchell, the designer of the Spitfire. He himself played Mitchell, with David Niven taking the part of a fictional character called Wing Commander Crisp, who was a composite portrait of the test pilots who nursed and developed the Spitfire from its cradle, including the great Jeffrey Quill. All the important roles were taken by professional actors, however Howard felt that a little authenticity might rub off if a few Battle of Britain pilots were to play themselves. That, at least, was the official reason, though it proved to

be a piece of misplaced optimism and the actors came over far more convincingly than the genuine airmen. Howard's main concern was his budget, I suspect. Serving officers could not be paid.

"I was one of the fortunate half-dozen or so pilots whose names came out of the hat, and the whole enterprise was the greatest fun. Our roles called for little more demanding than lolling about in our flying clothes in a fake dispersal hut, going outside to look sombrely skywards from time to time, and delivering such daft lines as, 'Good luck—they'll need it,' to cue in stock shots of Spitfires and German bombers flying overhead. Most of the stock shots in Second World War flying films include rapid frames of the interior of a Spitfire cockpit with a close-up of the pilot's gloved hand holding the control column, thumb on gun button in readiness for the decisive burst, before the montage cuts away to an external shot of an enemy aircraft being brought to grief. The same stock shots are wheeled out regularly to support documentary and news items to do with the battles in the air of the Second World War. That thumb on the firing button, which must have been seen by more people than any thumb in history, is my claim to immortality. Reader, the thumb is mine.

"The film-making episode with Leslie Howard was a joyous interlude and a marvellous break from airfield routine. I hardly remember when I enjoyed myself more. To make up for the fact that we were giving our services free, we were put up at the Savoy Hotel for the three or four weeks we were needed. What was more, a PR representative of the production com-

pany was placed at our disposal to make sure we were properly looked after once each day's shooting was over. This meant an endless nightly round of London's best clubs and pubs, no expense spared, all paid for by an apparently grateful movie management.

"The word soon got round, as words do. It was remarkable how many off-duty members of various RAF units 'just happened' to be passing the Savoy whenever we were setting forth on our nightly expeditions. To ask friends to join us was a handy, painless way of returning earlier hospitality. The clubs we frequented most were the Four Hundred, whose dimly lit dance floor and seductive music were excellent backdrops for young men in uniform trying their luck with the 'I'm off to war tomorrow and may never come back' routine; and the down-to-earth Bag o' Nails, a far more raucous affair where serious boozing took priority. We would then all finish up back at the Savoy, revelling in the music of the great Carroll Gibbons, that lovely man, pianist and bandleader, and his Savoy Orpheans. All in all, we fared extraordinarily well, and considered ourselves to be more than amply rewarded for our meagre, amateur contribution to the film."

". . . no other aircraft has ever had quite the charisma, almost a mystical aura, that the name Spitfire can evoke in both young and old. If you sit near to the Spitfires on display in the Royal Air Force Museum at Hendon, you will see large numbers of small boys pass by the other famous aircraft nearby with only a cursory glance, but they invariably stop to gaze at the

slender lines of R.J. Mitchell's masterpiece."
– Jeffrey Quill

"The aircraft was part of you and, when frightened, either in testing or in combat, I think one used to talk to one's Spitfire, and you may be equally sure that it used to answer."
– Wing Commander Patrick "Paddy" Barthropp

What leaps to mind when one thinks of Britain? In the 2006 Great British Design Quest survey by the BBC's Culture Show, to discover and rank the most iconic British things, R.J. Mitchell's Supermarine Spitfire ranked third among the twenty-five finalists. It was surpassed only by Concorde and the Underground Map. For Britons—whether aviation enthusiasts or ordinary folk—the charisma of the Spitfire endures today. So do more than fifty flying examples of the beautiful fighter. They continue to be lovingly cared for by their proud owners and many of these aircraft appear frequently in air shows staged in the United Kingdom, United States, Europe and elsewhere. Crowds flock to see the Spitfire and to hear the unmistakable growl of its engine more than seventy years after its maiden flight.

Spectators at such shows invariably engage in friendly debate about which aircraft was the greatest piston-engine fighter of all time. While the Spitfire has legions of devotees to defend its claim to the title, there are many others who would make a case for the superb North American P-51 Mustang, the Focke-Wulf Fw-190, and others. It is generally accepted that "greatest", in this context, breaks down into categories by function. The Mustang, for example, excelled in the key role of long-range escort for the American heavy bombers, while the Fw-190 was a stunningly effective interceptor. The Spitfire was designed specifically as a relatively short-range defensive fighter. It was intended to protect Britain from German attack and to destroy as much of the German Air Force in the air as possible. That accomplished, it lent itself brilliantly to further development. As the only Allied fighter operational from the first to the last day of the Second World War, it was produced in two dozen different versions and in a quantity surpassing 22,000.

Much of the magnificent reputation of the Spitfire rests on the public perception that it won the Battle of Britain. However, as has been pointed out by historians over the years, the majority of German aeroplanes destroyed in that battle were actually disposed of by its workhorse partner, the Hawker Hurricane. But it was the gorgeous Mitchell design that grabbed the headlines and captured the imagination of the public.

Flight Journal Editor-in-Chief Budd Davisson reflected: ". . . the Spitfire is so well-behaved compared with most other fighters. Its overall persona is better manicured, almost prissy—sort of like a heavyweight boxer who drinks tea with his pinky in the air just before breaking some thug's nose. It's difficult to describe. Whatever it is about the Spitfire that I can't put my finger on, I am sure that every single pilot who ever flew one has come away with the same sensa-

ORDE
29 Dec 1940

"SAILOR"

tions. This well-mannered but fire-breathing combination is also what made the plane such a winner in the hands of young pilots who, during the early days of the War, had barely graduated flight school. One month, they were wafting above meadows in a Tiger Moth, and the next, they were dueling with Messerschmitts and Heinkels at 25,000 feet.

"Clichés are clichés because they are so true and are often repeated, and we're going to repeat one yet again: the Spitfire is England. You can't say any given aircraft is the U.S., but when you look at a Spitfire, you forget all of its comrades in arms—the Hurricanes, Typhoons, etc—and simply see a nation at its finest." And Ervin Miller, who flew Spitfires with No 133 (Eagle) Squadron and later with the 336th Fighter Squadron, 4th Fighter Group, Eighth USAAF: "Even now, many years after I flew them on operations, the mere sound or sight of a Spitfire brings me a deep feeling of nostalgia and many pleasant memories. She was such a gentle little airplane, without a trace of viciousness. She was a dream to handle in the air. I feel genuinely sorry for the modern fighter pilot who has never had a chance to get his hands on a Spitfire. He will never know what real flying is like."

"I want to build a fighter, the fastest and deadliest aeroplane in the world. It's got to do 400 miles an hour, turn on a sixpence, climb 10,000 feet in a few minutes, dive at 500 without the wings coming off, carry eight machine-guns . . ."
– R.J. Mitchell (Leslie Howard) in *The First of the Few*

TEN of MY RULES for AIR FIGHTING.

1 Wait until you see the whites of his eyes.
   Fire short bursts of 1 to 2 seconds and only when your sights are definitely 'ON'.

2 Whilst shooting think of nothing else, brace the whole of the body; have both hands on the stick, concentrate on your ring sight.

3 Always keep a sharp lookout. "Keep your finger out"!

4 Height gives You the initiative.

5 Always turn and face the attack.

6 Make your decisions promptly. It is better to act quickly even though your tactics are not the best.

7 Never fly straight and level for more than 30 seconds in the combat area.

8 When diving to attack always leave a proportion of your formation above to act as top guard.

9 INITIATIVE, AGGRESSION, AIR DISCIPLINE, and TEAM WORK are words that MEAN something in Air Fighting.

10 Go in quickly – Punch hard – Get out!

IN THE BATTLE OF BRITAIN MOST RAF OPERATIONS HUTS CONTAINED A SMALL WALL POSTER WRITTEN BY THE FIGHTER LEADER ADOLPH 'SAILOR' MALAN. HIS RULES OF AERIAL COMBAT SAVED THE LIVES OF MANY RAF FIGHTER PILOTS IN THE SECOND WORLD WAR.

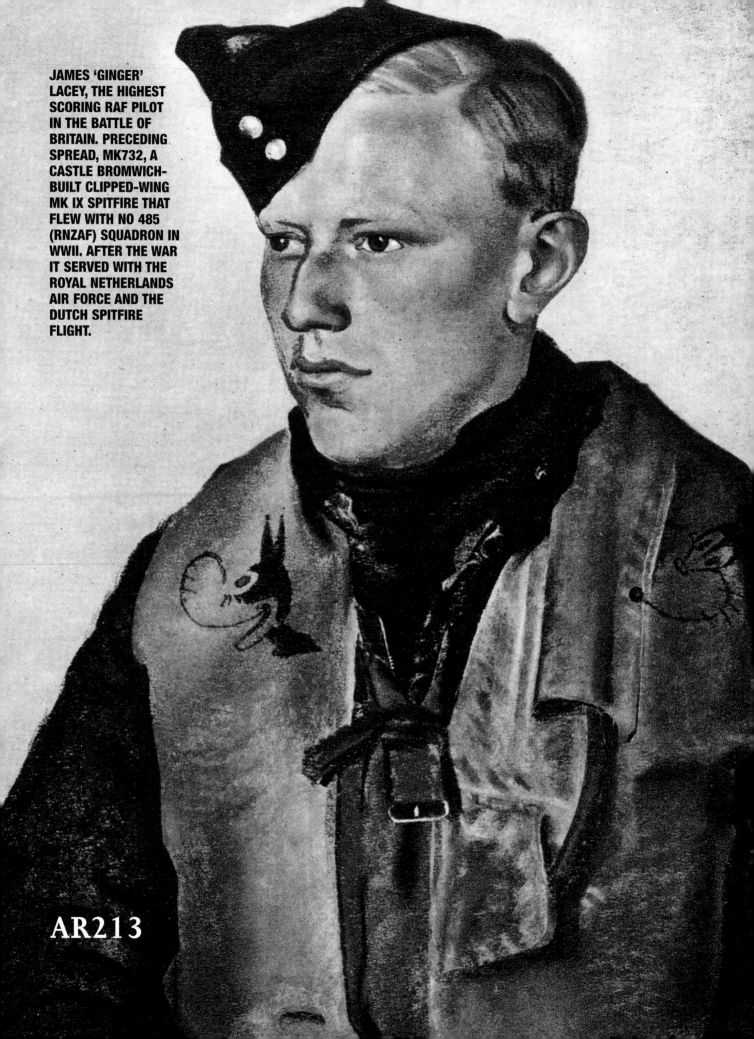

JAMES 'GINGER' LACEY, THE HIGHEST SCORING RAF PILOT IN THE BATTLE OF BRITAIN. PRECEDING SPREAD, MK732, A CASTLE BROMWICH-BUILT CLIPPED-WING MK IX SPITFIRE THAT FLEW WITH NO 485 (RNZAF) SQUADRON IN WWII. AFTER THE WAR IT SERVED WITH THE ROYAL NETHERLANDS AIR FORCE AND THE DUTCH SPITFIRE FLIGHT.

AR213

"Rumble thy bellyful! Spit, fire! Spout, rain!"
– *King Lear*, Act three, Scene two, by William Shakespeare

Ordered in July 1940 at the start of the Battle of Britain, Air Ministry Contract number 124305/40/C.23, Vickers-Supermarine Spitfire Mk Ia AR213 was built by Westland Aircraft of Yeovil, Somerset, and delivered to the Royal Air Force in February 1941. Westland was paid £6,600 for the aeroplane which was the second unit of the first batch of 300 Spitfires built by Westland. It was among the last Mk Is constructed and was sent to the storage depot, Number 12 Maintenance Unit at Kirkbride, Cumbria. A week after arriving at 12 MU, the Spitfire was sent to Number 57 Operational Training Unit at Hawarden near Chester where it was exposed to the hazards of flying training in the hands of many would-be fighter pilots. For part of its time at Hawarden, AR213 wore red, white and blue stripes on the length of its engine cowlings. A sister Spitfire, AR212, was serving with 57 OTU at the same time and its nose was painted yellow. They were both used as "bounce" aircraft to teach the student pilots to keep their eyes open, always know what is going on around them (known today as "situational awareness") and stay alert to keep from being shot down in combat.

James Harry "Ginger" Lacey, a Sergeant Pilot who had shot down more enemy aircraft during the Battle of Britain than any other RAF fighter pilot, was promoted to Flight Lieutenant after the Battle. On August 18th 1941 he was posted from 501

Squadron to 57 OTU for a so-called "rest" as an instructor. At No 57, he took up residence in Lord Gladstone's Hawarden Castle.

Before the Battles of France and Britain, Lacey had been a fresh-faced, tousle-haired kid—young, eager and relatively inexperienced. By the time he arrived at the OTU he had more than 300 operational hours in his logbook and eighteen German kills in his account. When, on September 13th 1940, a number of enemy bombers appeared over central London and one of them managed to drop its load on Buckingham Palace, Lacey caught up with the Heinkel and promptly shot it down while taking considerable return fire from the fleeing bomber. His own aircraft was badly damaged and he was forced to bale out, suffering minor injuries and burns. By the time he arrived at the OTU, his features had hardened and his boyish enthusiasm had mellowed into a steely confidence. He would finish the war with a total of twenty-eight enemy aircraft destroyed and nine damaged.

While instructing at Hawarden, Lacey flew the Mk Ia Spitfire AR213, coded JZ-E, as his personal aircraft. Instructing proved something less than restful in the company of fledgling pilots of little experience, varying skills and often questionable judgement. "I didn't give very much dual instruction: I'd been paid for doing that in peacetime, but in wartime I didn't think that the extra risk warranted it!" Ginger marked time until the six-month stint was over. The time had passed slowly for him as he found the teaching job boring and longed to be back on operations. Finally, he was posted again, to 602

Squadron at Kenley, south of London. In a letter of goodbye, Lacey's commanding officer at 57 OTU, Group Captain David Atcherley, wrote: "You may, or may not have guessed it, I don't know, but it was apparent to me from the word 'go' that your Flight was the best in the OTU. You have not only turned out good pupils, but you have also turned out good Instructors, and it was evident too that it was a happy concern from top to bottom. Well you have left your mark, and we will see to it that the standard is kept. I am not very hopeful of securing any official commendation for you, and anyway it is not a thing to discuss, nevertheless I would like you to know that if it lies within my power you shall have some recognition!

"All of us wish you luck and are certain of the successes in front of you. You can reflect, if the subject interests you, on the probability or otherwise of having done more good for the general cause by your work here, than hundreds of others have; even in sweeps. At least that is what I think."

Another high-achieving fighter pilot who almost certainly flew AR213 at some point was Bob Doe, who also distinguished himself in the Battle of Britain. Doe was posted to 57 OTU on October 22nd 1941 and served there as an instructor until June 9th 1943, when he was sent to the Fighter Leaders School, at RAF Milfield, Northumberland, before being returned to operational duties. While at Hawarden he was given command of the gunnery school, teaching new pilots how to fly Spitfires and how to shoot in air combat. He recalled two young men who came through the training during his

THE MK 1A AR213 OVER THE HEDGEROWS OF MIDDLE ENGLAND; RIGHT: AR213 BEING MADE READY BY SIMPSON'S AERO SERVICES AT RAF HENLOW FOR ITS ROLE IN THE BATTLE OF BRITAIN MOVIE IN 1968.

time on 57 and went on to become very high achievers themselves, Don Gentile and George Beurling. Don Gentile entered the RAF via the Royal Canadian Air Force before the United States had come into the war. He became a member of 133 (Eagle) Squadron, an American volunteer outfit, along with numbers 71 and 121 (Eagle) Squadrons, all of which were ultimately to form the famous 4th Fighter Group of the USAAF when the Eagle squadrons were transferred into the American air force. With more than 1,000 enemy aircraft destroyed, the 4th FG was the highest scoring American fighter group and Major Gentile was credited with twenty-eight of those victories. The Fourth brought their RAF Spitfires with them when they began operations from Debden, Essex, switching later to P-47 Thunderbolts and finally

to P-51 Mustangs. At the end of hostilities Gentile returned to his native Ohio to join the test pilot group at Wright Field. He was killed on January 28th 1951 in the crash of a Lockheed T-33 jet trainer. In his days at Debden, Don Gentile had formed a friendship and partnership with another American pilot, Johnny Godfrey, whose extraordinarily keen eyesight, marksmanship and flying skills he admired. As leader and wingman respectively, Gentile and Godfrey accounted for fifty-eight German planes destroyed and were lauded for their efforts by Winston Churchill and General Dwight Eisenhower.

It is believed that the Canadian Spitfire pilot George "Screwball" Beurling flew AR213 during his time at Hawarden too. On completion of his training at 57 OTU, Beurling served a brief time with 403

Squadron in which he downed two enemy aircraft, before being posted to Malta and 249 Squadron. In just three days with the squadron he destroyed eight German and Italian aircraft. Screwball seemed to be his favourite word and reference for many people in his circle, and soon it had become their nickname for him. His score of e/a downed reached seventeen and he was commissioned an officer. He finished the war with 403 Squadron back in England and thirty-one confirmed victories. Like Gentile, Beurling would die in an aircraft accident. In 1948, after a period in civil aviation, he became bored with the work and decided to join the young Israeli Air Force. He was killed near Rome while on his way to the Middle East.

With Leslie Roberts, George Beurling wrote of his experiences in a

fine book called *Malta Spitfire*: "The instructors at the Operational Training Unit, to which we reported for the final stage of pre-combat flying, were an all-star aggregation, mostly members of the few to whom the many owe so much—the birds who flew the Battle of Britain. Top-notcher, beyond question, was Flight Lieutenant Ginger Lacey, D.F.M. and Bar, with twenty-five German Aircraft Destroyed to his credit as a sergeant pilot from the Big Blitz and the pre-Dunkirk days across the Channel.

"I'd come down to OTU from seven days' leave in London and for the first time I can remember I was plain, red-headed mad with the whole damned German race. I'd seen things during that week, the sort of things I'd looked at before in pictures, but not close at hand in the flesh.

"One of those days I was in Tottenham Court Road when the Huns lifted the tailboard and let go everything they had, and there'd been a little bit of a girl sitting on the curb, playing with a doll. A policeman and I had seen her at the same time and had run over to her, because she didn't even seem to know the raid was going on. As I got there the bobby was saying, 'Better run along, baby. It's too dangerous to be playing here!' Then we both saw. The child was alive all right. But she was stunned, or in a stupor, just sitting there looking at her unhurt doll—her own right arm blown clean off at the shoulder!

"For me, from that day forward, I say God damn the Germans, the whole kit and caboodle of those who march with their bastard leaders and those who accept such leadership because they haven't the guts to die bucking it! That's tough talk. Roberts says maybe I'm oversimplifying the problem. Well, maybe. I kind of have a hunch it's the way the Russkies feel too, and a lot of ordinary guys in England. *And* the Maltese. There's only 250,000 of them. But believe me they're not feeling friendly to the Huns, or to the Eyeties either. Would you, if your home town had been knocked cockeyed, and all your kids were either dead or crippled? Hell, I better go hire a hall! But that's the way I felt, after seeing that little baby and her doll . . . I began to feel ashamed that up to now I'd been thinking of this air war as nothing but a great adventure for those who can fly.

"You can't afford to be one of these hot-blooded haters in this game. Many a good guy I know has gone west because he felt so violent about Germans and so damned anxious to pot the one in his sights that he for-

AIR COMMODORE ALLEN WHEELER FLYING THE MK 1A AR213 OVER ENGLAND IN AUGUST 1972.

got about the Hun on his own tail. It's a cold-blooded business. And it's a cold-blooded job we've got to do on the swine.

"That's the mood I was in when I taxied out to the Operational Training Unit from the station and seven days in blitzed London. Give me a Spitfire! Let me get going! Then I got down to the job of making a combat flier, a fighter pilot, out of Beurling. There was plenty to do, and plenty to learn.

"The long and short of the fighter pilot's business sum up about like this. There are three factors: flying, shooting, and the physical fitness, plus eyesight, of the pilot. Take the last one out, simply by saying it's vital. You need the kind of fitness which makes you so keen you just live to fly. As to flying and gunnery, let me put it this way: You can be the best precision pilot in the world, but if you don't know how to shoot, you won't get Jerries; and you can be the best shot above-ground, but if you can't fly like Billy-be-damned, then Jerry will knock you off. If you skid or slip around the sky, then every burst will go wide of the mark. So it's flying and shooting together that do the trick. Only when the two have merged and become one co-ordinated instinct can you expect to shoot down more than the odd accidentally-hit enemy.

"The minute you slide into the cockpit of a Spitfire for the first time you realise you've acquired an office in which you're going to be an extremely busy fellow. Exactly what non-flying people think a fighter pilot's job consists of, I don't know, but from what I've picked up in ordinary conversation I sort of get the idea the general opinion is that you simply slide into a comfortable, if slightly cramped seat, switch on the engine, and fly away. After that you throw the thing around the sky until you get your sights on a German. Then you shoot the German and come home. Maybe people don't make it quite so simple as that in their minds, but that would seem to be the approach. If that was the case, nearly everybody would be flying Spitfires.

"When you graduate to Spitfires you acquire a compact instrument of sudden death, for the enemy if you can get him in range, for yourself if you don't keep your wits about you. Into thirty-six feet of space, wingtip to wingtip, are packed two cannon

and four machine-guns, usually harmonised to bring their most devastating fire to bear on a point three hundred yards ahead, plus shells for the cannon in tension drums and ammunition for the Browning machine-guns either belts or pans. Mostly your machine-gun ammunition will alternate—incendiary, tracer, ball, and armour-piercing; that for the cannon the same, without tracer. Guns are fired by three buttons on the control column. Number One lets go the machine-guns alone, Number Three the cannon alone, Number Two gives out with all you've got.

" . . . we'd come down to OTU to learn . . . and learn we did. Spits were all over the sky, in all kinds of weather, in formation, dogfighting, split-assing, doing all the things which, added together, comprise combat flying. All that was missing from the scene to make it complete, was the enemy. He'd be the next stop, the stop at the end of the line.

" . . . the common fund, the purpose of which was to set aside a sum of money, raised through individual misfortunes, for a big class blowout at the end of the course. Landing with a dead prop cost you a shilling, bending a prop or other slight damage cost five shillings, a down-wind landing seven-and-six. If you were careless and took off with your airscrew at coarse pitch, that would be ten shillings, please. Landing without undercarriage, or any major crackup arising from the pilot's own carelessness, cost the even pound. Adjudicator of fines was Ginger Lacey, our flight commander. By the end of the course Lacey held almost £40 in trust for the gang and everybody went to town. What a night that was!"

After serving at 57 OTU for about

twenty months, AR213 was transferred to 53 OTU at Llandow, Vale of Glamorgan, on February 20th 1943. There it suffered a Category Ac flying accident on April 19th. Repaired, she was returned to the unit on May 12th, after No. 53 had moved to Kirton-in-Lindsey, North Lincolnshire. She was involved in another mishap in the hands of an eager young pilot officer in November, keeping her out of the line-up for a further month. She remained with 53 OTU until August 17th 1944, when she was put in storage with No. 8 Maintenance Unit at Little Rissington, Gloucestershire. At the end of the war she became one of thousands of unwanted Spitfires awaiting scrapping and was struck off charge on November 30th 1945.

In 1946, AR213 was bought by Group Captain Allen Wheeler, then head of the Experimental Flying Department, the Royal Aircraft Establishment, Farnborough, supposedly for £25. Group Captain Wheeler purchased the aeroplane from the Air Ministry, collecting it from the Maintenance Unit at High Ercoll and registering it on October 2nd 1946 as civil registration G-AIST. According to him, the Spitfire was then "built up" for static display at RAF Abingdon, Oxfordshire, by arrangement with the then Abingdon station commander, Group Captain Fred Sowrey. At the time that he bought the Spitfire, he also bought two new Packard-built Merlin engines, still in their original packing boxes. He paid £10 for each engine, including the boxes, and later sold each box for £10.

Wheeler's interest in buying AR213 was to enter it in the sponsored UK air races which were to resume in 1948 after a nine-year lull caused by the war. However, he decided against racing the plane in favour of acquiring another, later-mark Spit of greater power and racing potential. In 1949, he sent AR213 by road to Old Warden airfield, Bedfordshire, the home of the Shuttleworth Collection, where it was placed in storage. Richard Shuttleworth had founded his aeronautical and automotive collection at Old Warden in 1928. He had been killed in August 1940 in the crash of a Fairey Battle, and Wheeler, who had been Shuttleworth's best friend, subsequently worked with Shuttleworth's mother on the future development of the collection, hence his interest in warbird preservation in the post-war years. Group Captain Wheeler then bought the Mk Vb Spitfire AB910, which he *did* rebuild for competition purposes and registered it as G-AISU. That aircraft is now a part of the Battle of Britain Memorial Flight at RAF Coningsby, Lincolnshire.

Of particular interest in the wartime career of AB910 is an event on February 14th 1945. On that day Flight Lieutenant Tony Cooper, a flight instructor with 53 OTU at Hibaldstow, Lincolnshire, was leading a group of trainees in four Spitfires which were sitting on the side of the runway, waiting to take off, when he noticed a Spitfire coming in to land with a strange silhouette on the tail. The shape turned out to be ACW2 Margaret Horton, a WAAF who that day made an unscheduled flight as a passenger on the Spitfire piloted by Flight Lieutenant Neill Cox. Margaret described the experience: "The Spitfire incident was due to the rather slap-happy way the flight was run, due, I think, to the pressure of work rather than any negligence of the

NCOs, who worked all hours in the summer. On some Spitfire airfields, when the wind was above a certain force, the order 'Tails' went forth, which meant that every Spitfire had to have a mechanic sitting on the tail while it taxied to the runway, to prevent the aircraft from tipping onto its nose. Upon reaching the take-off point the pilot stopped for the mechanic to slip down before he took off. That was the theory, but my pilot had not heard the final order and took straight off upon reaching the runway with me still in place. I was sure I had reached my end when I found myself in the air, and as usually happens upon such occasions, ceased to be frightened, but was then possessed with the idea that I should be saved if I had faith, and so, it happened. I landed quite unshaken, though I am normally a coward for heights." She wanted it clearly understood that she was simply obeying orders in sitting on the tail, like all the other mechanics.

LAC George Thomas was a Senior Tow Target Operator who happened to be in the air at the time of the Horton incident: "I remember Margaret Horton well. I recall the look of fright on her face as she realised her predicament. We were going into the circuit, preparing to land, when we spotted her on the tailplane. We reported it and then formated as close as we dared. We were very pleased when she reached Terra Firma safely." Flight Lieutenant Bob Poynton was at Hibaldstow until March 1945 and was a flight engineer with an office near the Flying Control tower. He was notified of the emergency with the message, "An aircraft has taken off with a WAAF on the tail."

"The aircraft was doing the usual left-hand circuit at about 500 feet, with what sounded like full engine power. It was in a pronounced nose-up attitude and travelling slowly. Clearly, the pilot, Flt. Lt. Cox, was struggling to avoid stalling. It was quite frightening to see—like watching a high-wire circus act where there is no safety net—only worse: nail biting, breath-holding anxiety, for what seemed a very long time, and made all the more dramatic by the noise of the Merlin at full throttle. My recollection is that Flying Control acknowledged the pilot's reporting the tail-heavy problem, but did not tell him the cause. I think it fairly unlikely that he knew, or he would have said so. The Spitfire rear-view mirror was fairly small and had a limited field of view because it was flat. So, it would not be aligned to take in the lower fin and tailplane area where Margaret was desperately clinging on."

Frank Andrew drove the ambulance and the fire engine at Hibaldstow. On the day of the Horton incident the ambulance was parked next to the control tower so tower personnel could shout instructions down to the crew in an emergency. Thus Andrew was at the Spitfire as soon as it landed. His instruction was to take Margaret Horton to the Medical Section immediately, but she would not hear of it, insisting that she was all right and only needed a cup of tea. The medical staff was equally insistent and she reluctantly agreed to go, but only *after* her tea.

Finally, there is the account of Flt. Lt. Cox, the pilot of AB910, the Spitfire in the incident. He attributed the almost miraculous outcome due to two factors: that Margaret Horton remained calm and held on to the aircraft tail very firmly, and what he had learned about handling an aircraft with elevator problems when he had been shot down in combat and had landed in the sea. Flt. Lt. Cox was totally unaware, before and throughout the extraordinary flight, that the WAAF had mounted the tail of his Spitfire. On take-off the aircraft went into a steep climb and Cox was unable to move the control column. He tried using both hands and then, recalling his earlier experience, resorted to putting his knee behind the column and pushing, which brought the nose down. The control problem had been caused by Horton tightly gripping the elevator and pulling it upwards. Flying Control did not contact Cox during the flight, but he called them briefly to say that he was having a problem with the elevators and would be making an emergency landing. He did one circuit at an altitude of 600 feet and prepared to land. As he was losing height the aircraft suddenly went into a steep dive. He reacted well again and made a near-perfect landing. After landing, Cox was still unaware of Horton's presence as she had jumped off the tail and run back up the runway before the aircraft had come to a stop, to recover her beret which had come off during the landing. While in flight Cox had considered baling out, but with insufficient altitude and the likelihood of the Spitfire plunging into the nearby village, he elected to land. There was no damage to the aircraft apart from what he believed was some nail varnish where Horton had gripped the tail. Later, when summoned to explain the incident to his commanding officer, Cox pointed out that there had been a number of other Spitfires taxiing around the perimeter track without ground staff sitting on the tailplanes.

With his recent promotion to Air Commodore in May 1963, Allen Wheeler was posted to RAF Abingdon, Oxfordshire, and took AR213 with him, keeping it in storage there until 1967. It was then that Group Captain Hamish Mahaddie, in the employ of the *Battle of Britain* film producers, made his worldwide survey of Spitfires, Hurricanes, Heinkels and Messerschmitts for possible use in the movie. Mahaddie hired AR213 from Wheeler and the Mk Ia was moved from Abingdon to 71 Maintenance Unit at RAF Henlow, Bedfordshire, on June 12th to be renovated for its film role. AR213 was one of the group of nineteen Spitfires and three Hurricanes that were made

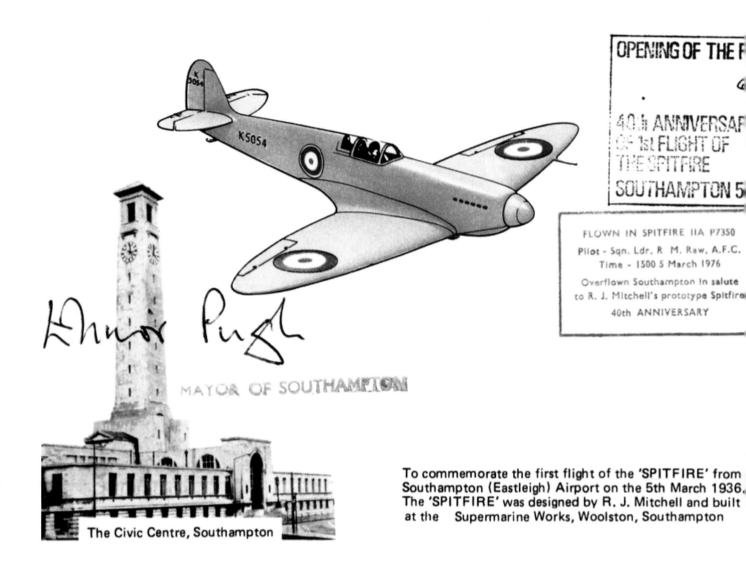

OPENING OF THE F[...]

40.h ANNIVERSAR[...]
OF 1st FLIGHT OF
THE SPITFIRE
SOUTHAMPTON 5[...]

FLOWN IN SPITFIRE IIA P7350
Pilot - Sqn. Ldr. R M. Raw, A.F.C.
Time - 1500 S March 1976
Overflown Southampton in salute
to R. J. Mitchell's prototype Spitfire
40th ANNIVERSARY

MAYOR OF SOUTHAMPTON

The Civic Centre, Southampton

To commemorate the first flight of the 'SPITFIRE' from
Southampton (Eastleigh) Airport on the 5th March 1936,
The 'SPITFIRE' was designed by R. J. Mitchell and built
at the   Supermarine Works, Woolston, Southampton

available to the film-makers.

In order to return these elderly aircraft to a high airworthy standard, and maintain them in that condition, the producers hired Simpson's Aero Services, then based at Elstree Airfield, Hertfordshire, a company renowned for its expertise on Merlin-engined aircraft. Simpson's did much of the work on the movie aircraft at the Henlow facility.

At Henlow, AR213 was fitted with a Merlin 35 engine supplied by Jersey Aviation, Channel Islands, as well as a four-bladed propeller. Air Commodore

Wheeler noted in correspondence following completion of the filming that the aeroplane was returned to him in superb condition. He later moved the Spitfire to Wycombe Air Park, the former RAF Booker, in Buckinghamshire, where he flew it occasionally and kept it in the care of Doug Bianchi's Personal Plane Services. On May 5th 1974, Wheeler sold the aeroplane to The Honourable Patrick Lindsay, second son of the 28th Earl of Crawford and 11th Earl of Balcarres.

Lindsay lived life very much in the

fast lane, constantly pursuing his passions, skiing, travelling, motor racing and piloting his own aircraft, having learned to fly in the RAF. He first raced in the 1950s at the Goodwood circuit, part of the wartime fighter station known as RAF Westhampnett, West Sussex. That early effort ended prematurely when his racing car caught fire. He survived more than his share of shunts, including one when he suffered several broken bones at Thruxton where his Maserati 250F crashed when the throttle jammed open. It took forty

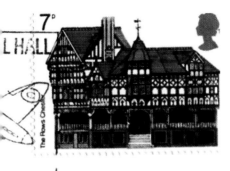

R. J. Mitchell Committee
The R. J. Mitchell Hall
Kingsbridge Lane
Southampton, SO1 0GB

minutes to cut him from the wreck. His personal air force was composed of a 1932 Moraine Saulnier, a WWI SE5a, a replica Sopwith Triplane, a Thirties Hawker Fury, a WW2 Westland Lysander, a Feisler Storch, and the Mk Ia Spitfire, AR213.

Alongside his special interests and his work at Christie's where he was a director, Lindsay was known as a great champion of the arts in Britain, and his charitable fund raising helped to keep many fine works of art in the country.

Lindsay continued to keep the

Spitfire at Booker with Personal Plane Services. Bianchi's son, Tony, former British aerobatic team member, manager and later owner of PPS remembers: "When Patrick Lindsay owned AR213, he would come out from Christie's in London. He would arrive in his D-Type Jaguar and his flying jacket. The flying jacket would come off when he got out of the Jaguar and he would sit in the aeroplane in a suit and braces, a tie and a flat cap. Patrick was a unique chap; a connoisseur of everything. No matter what it was, if it was collectible he

"IT IS NOT GOOD ENOUGH TO FOLLOW CONVENTIONAL METHODS OF DESIGN. IT IS ESSENTIAL TO INVENT AND INVOLVE NEW METHODS AND NEW IDEAS." SO SAID REGINALD MITCHELL, THE MAN BEHIND THE SPITFIRE. HIS INSPIRATION, GENIUS AND DETERMINATION, TOGETHER WITH THE BRILLIANT DESIGN TEAM HE HAD ASSEMBLED AT SUPERMARINE, LED TO THE CREATION OF THE FABULOUS LITTLE FIGHTER PLANE.

"WITH SUCH A RELATIVELY SMALL NUMBER ACTUALLY EXPERIENCING AERIAL WARFARE, IT IS DIFFICULT TO FIND FIRST-HAND ACCOUNTS. IN TOO MANY FAMOUS ACTIONS, PARTICULARLY TOWARDS THE BEGINNING OF THE WAR, HARDLY ANYBODY SURVIVED TO TELL THE PERSONAL STORY. AND IN ANY CASE, YOU CANNOT FIT A WAR CORRESPONDENT INTO A SPITFIRE."
—GAVIN LYALL, AUTHOR AND AVIATOR

loved it and he needed to own it. His whole life was having nice things around him. When I started motor racing I met him at various races and knew of his interest in old aeroplanes. He learned to fly in the Fifties with a university air squadron at Oxford. Then he gave up flying until the Sixties, when he turned up at PPS and booked an appointment with my father. They became extremely good friends. He wanted an unusual aeroplane; wanted to renew his license and have something wonderful, collectible and unusual. My father got a Moraine 230 in France for him. It was a nice, high-wing, parasol fighter trainer which Patrick flew in addition to the Spitfire." Patrick Lindsay died in January 1986 at the early age of 57. At the funeral in Hungerford, Tony Bianchi performed a farewell flypast in the Spitfire. Two months later, in honour of the fiftieth anniversary of the first flight of the prototype Spitfire, Tony flew AR213 in a display over Eastleigh airfield near Southampton, where that historic event had taken place. He then continued on with another flypast over the sites of the old Woolston and Itchen Supermarine works.

R. J. MITCHELL

LEFT TO RIGHT: MUTT SUMMERS, H. J. PAYN, R. J. MITCHELL, S. SCOTT-HALL, AND JEFFREY QUILL, AFTER THE MAIDEN FLIGHT OF K5054, THE SPITFIRE PROTOTYPE.

The little firm that began building what it called "water planes", or "boats that fly", at Southampton in October 1913 was called Supermarine. It was established by an eccentric aviation pioneer, inventor and property developer, Noel Pemberton-Billing. He and his partner, boating enthusiast Hubert Scott-Paine, took a liking to a disused coal yard on a piece of the River Itchen frontage between the old Ferry Yard and Floating Bridge in the Woolston area of Southampton and made it the site for their new aircraft works.

Supermarine became well-known and highly regarded in the aviation community for the quality and performance of their products, so much so that in 1928 they were taken over by Vickers and became Vickers (Aviation) Ltd, Supermarine Works, trading under that name for ten years. The heavy-weight aircraft makers, Vickers-Armstrong Limited, eventually took over both Supermarine Aviation Works and Vickers (Aviation) Limited, of Weybridge, to become the parent company in the manufacture of the Spitfire fighter plane.

Supermarine Aviation was self-styled and largely independent and managed to function with relative autonomy for half a century until it ceased trading in 1963. Through all that time it was referred to by its employees and most other people as "Supermarines".

Reginald Joseph Mitchell was "Reg" to his friends at Hanley High School in Stoke-on-Trent, in the English Midlands. They all knew him to be "mad about aeroplanes", according to his son, Dr Gordon Mitchell. Reg was born in Stoke on May 20th, 1895. His father, Herbert, was a headmaster in Longton near the city centre. As soon as they could manage it, the family moved to a cottage away from the industrial grime and smoke of the pottery town. Although it was called Victoria Cottage, the Mitchell home was large and comfortable. It easily accommodated Mr and Mrs Mitchell and their five children. Reg and his younger brother Eric spent much of their free time building model aeroplanes, not from kits as many boys do today, but from scratch, creating their own unique designs with wood, paper and glue.

Mitchell's interest in flying and all things aeronautical led him to keep racing pigeons for a while, sending them to France to fly in "homing" races. Reg was a strong leader type and highly determined. In 1911, when he was sixteen, he decided to leave high school in order to join the locomotive firm of Kerr, Stuart and Company as an engineering apprentice. The work and the environment were filthy and oppressive and Reg was most unhappy. He said as much to his father who told the boy, "You will keep going, my lad, and you will like it!" He went and did not like it, but, in time, was able to move on from the shops to the drawing office, which he did like. While there, he attended night classes in mechanics, engineering drawing and higher mathematics, having decided to make a career as an engineer. Reg did well at Kerr, Stuart, but in 1916, with the First World War in its second year, he tried to enlist in the forces. Fortunately, he was turned away twice and told that his engineering skills were of greater use in the civilian world. The next year he applied for a position as personal assistant to Hubert Scott-Paine, at Supermarine in Southampton. He was invited for an interview and was offered the job. He wanted it very much, but was reluctant to leave Stoke and Florence Dayson, the girl he was courting, and move so far away. He accepted the position, however, and was off to join his new employer.

In his eagerness to learn all he could about the company, its products, and aviation in general, Reg worked extremely hard, and his zeal and capabilities were quickly recognised by Scott-Paine who, within a year, made Mitchell assistant to the works manager. With the promotion in hand Reg returned to Stoke for a brief visit during which he and Florence were married. They rented a house in Bitterne Park, Southampton, and Reg bought a motorcycle and sidecar for his daily commute to Woolston. He continued to work hard and to impress his colleagues in the design team of only six draughtsmen and a secretary.

With the departure of the firm's chief designer, F.J. Hargreaves, in 1919, Reg was given the job. The following year, at age twenty-five, he was made chief engineer, a position he held until Supermarine was taken over by Vickers Ltd. He was subsequently appointed director and chief designer, the role in which he served until his tragic death in 1937. The ownership and top management of Supermarine had changed by the end of the Great War. In 1916, Pemberton-Billing had relinquished sole management to Scott-Paine who, in 1923, sold the company to another member of the firm, Squadron Commander James Bird. By

1919, the well-liked and greatly respected Mitchell had become affectionately known there as "R.J."

In November 1918, Supermarine moved into the design and construction of flying boats for civilian use. In that period the ambitious Scott-Paine wanted to win the coveted Schneider Trophy for Britain, a feat that would certainly enhance the prospects of the firm. The award had been established in 1913 by Jacques Schneider, the heir to a French armaments fortune. Schneider championed seaplane development in order to take advantage of the vast water-covered areas of the earth's surface as airports. The trophy challenge was to pit teams from the United States, Britain, France and Italy in a race over a measured course to determine the fastest seaplane among the competitors. At first, the race was held annually and later biennially, with the winning country stag-

ing the next event. During the war years the contest was suspended. It was resumed in 1919.

The French won the initial race which was held at Monaco in April 1913 and the British won the 1914 event, the last until the post-war resumption. The winning pilot of the first Schneider race was Maurice Prevost, with an average flying speed of 45 mph. The British winner of the 1914 event was Howard Pixton in a Sopwith Schneider biplane with floats. He averaged 86.78 mph. The 1919 race was flown off the coast of Bournemouth and became a fiasco when none of the entrants satisfactorily completed the course and the event was declared null and void. There were no British entries in the 1920 and 1921 races, which were both won by Italy.

The rules for the Schneider Trophy competition stated that the race must be staged over water and that the air-

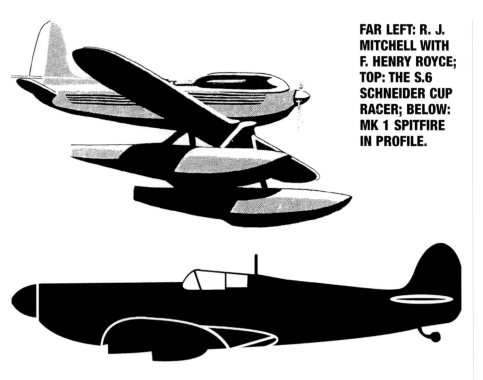

FAR LEFT: R. J. MITCHELL WITH F. HENRY ROYCE; TOP: THE S.6 SCHNEIDER CUP RACER; BELOW: MK 1 SPITFIRE IN PROFILE.

craft participating must be seaworthy. Each nation was limited to a maximum of three entries, to be sponsored by a governing body which, in the case of Britain, was the Royal Aero Club. If any nation accumulated three consecutive victories it would retain the Schneider Trophy. The race itself had to be flown over a distance of 350 km in a number of laps around a closed circuit course. Invariably, it proved to be a great crowd pleaser with the aircraft flying the tight course at extremely low altitude, making a lot of noise and providing the spectators with a magnificent view of the contest.

Scott-Paine and the Supermarine staff had plenty of motivation before the 1922 race. National pride dictated that the British must not allow the Italians to win a third successive event and thus retain the Trophy. Furthermore, the currency attached to such a winning record by a Supermarine entry would be immensely valuable to the

firm in its efforts to become a world-class flying-boat maker.

As there was no possibility of British Government funding for a Supermarine entry, the firm had to underwrite their advanced version of the Sea Lion flying boat design (the Sea Lion II) for the 1922 event to be held at Naples. In the race, Supermarine's test pilot, Henri Biard, managed a narrow victory over the Italian Macchi entry and proudly brought the trophy back to Britain.

Dr Gordon Mitchell: "The most extraordinary thing about Mitchell's success as an aircraft designer was that he was, to a large extent, self-taught. As has been seen, all his early training had been done on locomotives, and with his creative ability he might well have ended up designing the now commonplace high-speed trains. Instead, by some fortuitous circumstances, he found himself in an

aircraft factory and when he became chief designer, at the age of only twenty-four, a whole new world of exciting challenges opened up for him. He was always interested in speed, and his work on flying boats had given him the chance to design an entry for the Schneider Contest."

For roughly eleven years, it was R.J.'s task to design and oversee the development of the Supermarine racing seaplanes for the Schneider effort, a job made significantly more difficult and demanding by the complete absence of financial support from the Ramsay Macdonald government for that effort. In the Schneider competition work, Mitchell and the design team progressed through considerable refinement of the Sea Lion flying boat, to the famous S series of racers. The S.6B won the coveted trophy at Calshot in September 1931, in the last of three successive Supermarine victories, to retire the award in Britain.

Dr Mitchell: "Because of his high power of concentration he hated to be disturbed at work, and his team soon learnt to recognise the danger signal. When a man entered the office he stood just inside the door, waiting in a nervous silence. If Mitchell turned his head with a welcoming smile, then all was well. If, on the other hand, a red flush began to creep up the back of his neck, the man hastily fled before the storm broke. Mitchell could be brusque and quick tempered, and it was necessary to choose the right moment before interrupting him at work.

"But R.J. could also be very charming. He had a great sense of fun, and the smile which lit up his face when he was amused utterly transformed his usual grave expression. Away from

THE
SCHNEIDER TROPHY CONTEST, 1931
THE ROYAL AERO CLUB
OFFICIAL SOUVENIR PROGRAMME

Edited by SQUADRON-LEADER C. G. BURGE, O.B.E., A.R.A.S.I. R.A.F. (Retired)

OFFICIAL PROGRAMME MANAGERS
PRINTERS AND PUBLISHERS.

GALE & POLDEN LTD
AVIATION
DEPARTMENT

THE
JACQUES SCHNEIDER
MARITIME TROPHY
——"——
Presented in 1912 to the
AÉRO CLUB DE FRANCE
by M. Jacques Schneider for an
International Aviation
Competition under Rules
approved by the Federation
Aeronautique Internationale

2 AMEN CORNER. E.C.4.
ALDERSHOT
WELLINGTON WORKS
PORTSMOUTH
EDINBURGH ROAD

work he enjoyed many sports, and at such times he was a jolly, friendly companion. The one social event that he really enjoyed was the annual drawing office party, a 'men only' celebration usually held in one of the Southampton hotels. There were two opposing sides to Mitchell's character. He worked harder than any man at Supermarine, but when work was over he could throw himself wholeheartedly into a noisy party as though that was the only thing that mattered.

"Basically Mitchell was a shy man. He suffered from a slight stammer which worried him terribly and made him nervous when talking to strangers. But in the company of men he knew and liked he relaxed and the hesitation in his speech was forgotten."

When the Americans came to compete in the 1923 Schneider race at Cowes, Isle of Wight, they brought three impressive seaplanes; two were Curtiss CR-3s and the third a Wright NW-2. They also brought four U.S. Navy aviators eager to show the British and French teams how well their streamlined racers performed. One of the Americans, Lt. Rittenhouse, won with an impressive average speed of 177.38 mph, some twenty miles an hour faster than the British Sea Lion III entry had achieved.

For R.J. Mitchell, the defeat in 1923 was a significant lesson. He realised that the age of the racing flying boats was over. The amazing performance of the Curtiss racers pointed the way to creating an entirely new and special class of plane to help Supermarine regain Schneider glory for Britain and the company. He rededicated himself to that achievement and this would lead to the most important accom-

plishments of his career. Foreign victories in the recent Schneider contests forced the leading British aero-engine makers to reconsider their approaches to engine design, resulting in a new product from Rolls-Royce. The Kestrel would be used to power the Hawker Hart and Hawker Fury aircraft, and its evolution into the Kestrel V would ultimately be developed into the prototype PV-12, the famous Merlin, whose lineage would power a large portion of the Spitfire production.

So impressed and appreciative of R.J. Mitchell's genius and his great contribution to the firm were the bosses of Supermarine that, in December 1924, they offered him a ten-year employment contract which included a technical directorship.

The 1925 Supermarine entry in the Schneider competition, the S.4, was powered by a Napier Lion engine of 700 hp. In a test flight on the day before the Baltimore race, the aircraft was being flown by Henri Biard who got into trouble around the first marker. The horrified Mitchell watched from a rescue boat as the S.4 crashed, raising a huge plume of white spray. The force of the impact broke the back of the racing plane but Biard survived with no serious injury. He later said that he thought the accident had been caused by wing flutter during that first turn. Later investigation laid the cause on a stall which had possibly been brought on by aileron flutter. The American Army pilot, Lt. Jimmy Doolittle, won the race in a Curtiss R3C-2 at an average speed of 232.57 mph. The disappointing S.4 set-back challenged Mitchell to find ways to overcome its

deficiencies while retaining the basic design which he firmly believed to be sound. His design genius and his dogged determination would bring him ultimate triumph.

Along with the work on the Schneider racing planes, 1925 saw the completion and first flight of a new and significant armed military flying boat—the Southampton. Henri Biard flew the mahogany-hulled Mk I in March and the Air Ministry were sufficiently impressed with its performance and its potential for reconnaissance and anti-submarine patrol work to order it into quantity production. The Southampton crew consisted of two pilots, a wireless operator, a gunner, and a bomb-aimer. It cruised at 85 mph, had a top speed of 107 mph and a range of 680 miles. The efficiency of Mitchell and his design team was amply demonstrated as they propelled the Southampton from initial drawings through construction and testing in only seven and a half months. The impressive demonstration flights by Southamptons to various parts of the globe lent enormous credibility to the growing fans of long-distance flying boat travel. They saw many advantages in such aircraft over land planes: the ability to operate from water at destinations where airports did not yet exist and the comforting safety factor in being able, should trouble develop, to alight on the water. The Southampton gained world-wide fame, enhancing the growing Supermarine reputation for designing and building some of the best flying boats ever made.

There would be no British entry in the 1926 Schneider event, staged in

Virginia, as the Napier Lion engine being built for the new Supermarine S.5 entry was not ready in time for the race, which was won by the Italians. The S.5, a substantially improved aircraft over its predecessor, was finally ready for competition in the 1927 event, to be held at Venice. The aeroplane was almost entirely constructed of duralumin and featured a fully-braced wing and float structure. The wing was of all-wood construction and the plane was powered by a Napier Lion VII engine of 900 hp, tightly fitted into the minimalistic design of the cowl. The aircraft was smaller than the S.4 and the cockpit tiny. With little room on-board, part of the fuel was carried in the starboard float, the effect of which also helped to counteract the torque of the propeller.

In May 1927, the year of the last annual event, the Air Ministry finally agreed to finance the British entry for the Schneider race and the RAF took on the flying responsibility for Britain, organising a special unit for that purpose—the High Speed Flight. An S.5 in the hands of Flt. Lt. S.N. Webster won the race with an average speed of 281.66 mph.

It was largely due to the performance of the S.5s, as well as that of the Supermarine flying boats, that Vickers entered negotiations in the autumn of 1928 to buy the Woolston firm. It was clear that the Vickers people wanted R.J. Mitchell as the key part of the package, making that a prime condition of the deal. Mitchell's ten-year agreement with Supermarine meant that he would be obligated to his firm, and to whatever firm might acquire it, until the end of 1933. The

purchase was concluded in November 1928. By June 1931 the name had been changed to the Supermarine Aviation Works (Vickers) Ltd and in that same year a Scottish engineer, Sir Robert McLean, was appointed chairman of the Vickers aviation companies. It would be McLean who would make it possible for Mitchell to create the Spitfire.

The 1929 British Schneider entry saw the involvement of the Rolls-Royce aero-engine firm in the Supermarine S.6, an aircraft that Mitchell was designing around a new engine that Rolls had been asked to provide. Time was a critical factor, as of course, was cost. The British government agreed to support the nation's entries and both Supermarine and Gloster made plans to enter. Rolls-Royce chose to save some time by modifying one of their existing engines, the 825 hp Buzzard, which was essentially a scaled-up Kestrel. The result was the R-type. When made stronger and fitted with a double-sided centrifugal supercharger, its power output at sea level was increased to a tremendous 1,850 hp. But before turning away from the Napier Lion engine he had so long relied upon, Mitchell went to see Sir Henry Royce to talk about the new engine concept. Royce promised to deliver a powerplant of at least 1,500 hp with a development potential to at least 1,900 hp. Mitchell calculated that, with it, the S.6 would be capable of a 400 mph top speed, a level speed of 350 mph, a dive speed of 523 mph, and a rate of climb of 5,000 feet per minute. The S.6 would have to be a bit larger than the S.5, to accommodate the new engine. It was the start of a most fruitful partnership

between Rolls-Royce and Supermarine.

Dr Gordon Mitchell: "The S.6 was not an easy aircraft to fly. It needed quite a lot of force to move the ailerons, while the elevator control required a very light touch. Good cornering was essential for the Schneider circuits, and . . . that demanded special skills for, owing to the pull of centrifugal force, pilots temporarily 'blacked-out' on every sharp turn." On September 7th, the day of the 1929 race which was being staged on the Solent near Calshot, thousands of spectators crowded the best vantage points. It was estimated that a million and a half people had arrived to witness the event. Offshore, the aircraft carrier HMS *Argus* hosted two prominent observers, the Prince of Wales and the Prime Minister, Ramsay Macdonald. Flying Officer Dick Waghorn won the race in an S.6 with an average speed of 328.63 mph. With two consecutive victories Britain was poised to retain the Schneider Trophy outright if it could win the race in 1931. The 1929 victory had been a great coup for Mitchell and Supermarine. An appreciative management made it possible for R.J. to drive a Rolls-Royce car from then on. With his international design reputation made, he began to receive lucrative offers of employment from foreign aircraft makers—offers far in excess of his salary at Supermarine. As his son recalled: "If he had accepted any of them he could have become a very wealthy man, but even if he had wanted to, which he didn't, he was committed by his agreement of 1924, and that with Vickers in 1928, to remain at Supermarine at least until 1933." At the time of the 1931 Schneider

THE
WORLD'S FASTEST AIRCRAFT
319·57 M.P.H. AT CALSHOT, Nov.4TH.1928.
GAINING BRITISH
SPEED RECORD

FLIGHT LIEUT. D'ARCY GREIG, D.F.C., A.F.C., R.A.F.
WITH MODEL OF THE
SUPERMARINE NAPIER S.5. HIGH SPEED RACING SEAPLANE
WHICH HE PILOTED WHEN THE ABOVE RECORD WAS OBTAINED.
THE SAME MACHINE PILOTED BY FLIGHT LIEUT. S.N.WEBSTER, A.F.C. R.A.F. AT VENICE, 26TH SEPT. 1927.
GAINED FOR GREAT BRITAIN
THE INTERNATIONAL SCHNEIDER TROPHY
AT AN AVERAGE SPEED OF 281·51 M.P.H.
THE SUPERMARINE AVIATION WORKS, LTD.
PROPRIETORS—VICKERS (AVIATION) LIMITED.
SOUTHAMPTON. ENG.

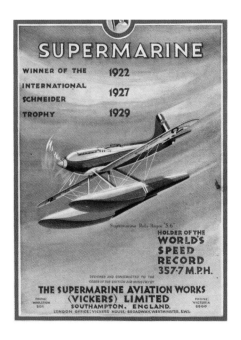

FLOWN FROM ROYAL AIR FORCE CONINGSBY
IN SPITFIRE MK. 19 PF 853 TO COMMEMORATE
THE 40TH ANNIVERSARY OF THE FIRST FLIGHT
OF THE SPITFIRE.

PILOT: SQN. LDR. M. RAW, AFC, RAF

FLT. TIME: 10 MINUTES

The Commandant
Eastleigh Airport
Southampton

race, Britain and much of the world were in the grip of economic depression. More than two million people were out of work in the British Isles and the Vickers group, along with most other companies, was struggling. There seemed no possibility of the government lending financial support for the British Schneider entries, despite the lobbying of the Royal Aero Club and the Society of British Aircraft Constructors. The goal of retiring the trophy with a third successive British win appeared hopelessly out of reach. In January 1931 the Ramsay Macdonald government announced that expenditure of public money in support of the contest could not be justified in light of the present financial situation. Debate raged in the nation's newspapers, but the government's position remained unchanged.

An entirely unexpected angel appeared in the form of Dame Fanny Lucy Houston, a philanthropist with a sizeable fortune inherited from her late husband, a shipping magnate. She donated £100,000, the cost of building two new racing planes for a British Schneider entry. She also took the opportunity to fire a blast at the Prime Minister: "I know I can confidently rely on the kindly help and co-operation of all who will rejoice if England wins." And to the London newspapers she commented: "We are not worms to be trampled under the heel of Socialism, but true Britons."

Mitchell would have the opportunity to build a faster, better racer in the S.6B. With only seven months remaining, there was no time to design and build a brand new plane (the S.7), so R.J. decided to refine the highly successful S.6, call it the S.6B and power it with the second generation Type R

engine that Henry Royce had promised in their initial meeting. Royce delivered on the promise with a 2,350 hp monster of enormous potential.

As it turned out, the 1931 Schneider competition was not a true race. The French withdrew their entry and, in August, one of the Italian pilots was killed in a crash while testing a new Macchi 72 over Lake Garda. Both France and Italy requested a postponement of the race, but the British Royal Aero Club would not consider it. The event went ahead as scheduled on September 13th with RAF Flt. Lt. J.W. Boothman setting an average speed over the course of 340.08 mph, and the Schneider Trophy went to Britain in perpetuity. The racing seaplanes created by R.J. Mitchell and his team at Supermarine had laid the groundwork for the Spitfire, the fighter that in less than a decade would prove so vital to the survival of Britain in the Second World War.

A year before the British retention of the Schneider Trophy, the Air Ministry put out a call for a new single seat monoplane day and night fighter capable of at least 195 mph, with exceptional manoeuvrability, considerable range, high initial rate of climb, a low landing speed, and good visibility for the pilot. The armament was to be four .303 Vickers machine-guns and the aircraft was to be fitted with a radio/telephone. The aeroplane makers Westland, Blackburn, Bristol, Gloster, and Supermarine all prepared designs for the specification. Their efforts were unsuccessful and the Supermarine aeroplane, the Type 224, was a great disappointment for Mitchell. In the end, the Air Ministry

had no choice but to adopt the Gloster Gladiator biplane for its next fighter.

However, Mitchell's design for the Type 224 was of far more value to the firm than it had seemed during its development. After the failure of the plane in testing, the Air Ministry tried to persuade Supermarine to replace the Goshawk engine then powering the Type 224, with a Napier Dagger engine. The Supermarine management rejected the idea as a waste of time. Mitchell, meanwhile, began to revise the design of the 224, incorporating many important refinements including a retractable undercarriage, and a much thinner wing. In time it began to resemble the first Spitfire.

In summer 1933, R.J. was taken seriously ill. His doctors diagnosed cancer and he was operated on at St Marks Hospital, London, in August. His doctors told him that there was a high risk of the cancer returning and, if so, it was unlikely that there would be anything more that they could do for him. He courageously committed the time he had left to his work. Dr Gordon Mitchell: "None of the people working closely with him realised the seriousness of his illness. Mitchell never wanted to talk about himself, only about getting the new fighter design right this time."

In November 1934, the chairman of Supermarine, Sir Robert McLean, met with his counterpart at Rolls-Royce, A.F. Sidgreaves, and they agreed that their two companies should privately finance the design and construction of a new fighter prototype. Rolls-Royce contributed £7,500 towards the design effort.

**THE PROTOTYPE SPITFIRE K5054;
BELOW: THE GRAVE OF MITCHELL,
EASTLEIGH NEAR SOUTHAMPTON.**

IN LOVING MEMORY OF
REGINALD J. MITCHELL, C.B.E.
AT REST 11TH JUNE 1937.
ALSO OF FLORENCE
HIS BELOVED WIFE
WHO PASSED AWAY 3RD JANUARY 1946

UNTIL THE DAY BREAK.

When they informed the Air Ministry of their intention, they made it clear that under no circumstances would any technical member of the Air Ministry be consulted or allowed to interfere in any way with the design of the new plane which McLean referred to as the "killer" fighter. The decision to build the plane, and his faith in Mitchell to create it, are to McLean's great credit. The Supermarine Board authorised and directed Mitchell to begin design work on the new fighter, an aeroplane that would be powered by the Rolls-Royce PV-12 (PV stood for Private Venture) with an ethylene glycol liquid cooling system. The plane was allocated the company Type number 300.

After many conversations with his chief aerodynamicist, Beverley Shenstone, Mitchell decided on the now famous elliptical shape for his wing. He wanted the wing to be as thin as possible, within the strength requirement, but it would have to be thick enough near the root to fit the retractable undercarriage and the machine-guns. Nearly all of the aeroplane would be of stressed skin construction, a process that, in the 1930s was both quite advanced and most difficult. It would have an armament of eight .303 machine-guns. Improved visibility for the pilot was assured through the addition of a sliding perspex canopy which also reduced drag. Chief Draughtsman Joe Smith was in charge of all the detail design for the new plane.

Early in 1936 the prototype fighter, serial number K5054, was approaching completion. As the time for the maiden flight grew near, so did the time to name it. Shrike, and Shrew, were among the suggested names. Sir Robert McLean, however, was adamant that it should be called Spitfire, a name that had already been used for the ill-fated Type 224 and was, therefore, not favoured by many in the Supermarine organisation. For some, it was a bad omen associated with failure and Mitchell himself thought it a "silly" name. Fortunately, McLean prevailed, to the enthusiastic approval of millions of Spitfire lovers ever since.

Most aviation historians have referred to the first flight of the Spitfire prototype as having taken place on March 5th 1936, but Jeffrey Quill, the great test pilot who took over from Joseph "Mutt" Summers as Chief Test Pilot for the Spitfire programme at Supermarine, refutes that in his book *Spitfire*: "Popular folklore has it that the first flight of the Spitfire was on 5 March 1936 but I flew Mutt to Eastleigh for the particular purpose of making that first flight on 6 March." More on this in the *Early Spitfire* chapter.

The headline of the *Southern Daily Echo*, the Southampton local paper, read: NEW SUPERMARINE FIGHTER FIVE MILES A MINUTE MONOPLANE HUSH-HUSH TRIALS AT SOTON. "Keen observers in and around Southampton have recently been interested in the high-speed performances of a remarkable plane which has made occasional flights from Eastleigh Airport. This machine is the very latest type of single-seater fighter, designed and built for the RAF by The Supermarine Aviation Works (Vickers) Ltd at their factory in Woolston.

"Produced amid great secrecy, the plane is one of the fastest of its category in the world. Like all Supermarine air-craft, the new fighter was designed by Mr R.J. Mitchell, CBE, director and chief designer of the firm, who designed every British winner of the Schneider Trophy since the war. Even the uninitiated have realised when watching the streamlined monoplane flash across the sky at five miles a minute (300 mph) and more, that here is a plane out of the ordinary."

The Spitfire prototype had to pass Air Ministry flight testing at Martlesham Heath in Suffolk. Flt. Lt. H.E. Jones was assigned to make an initial assessment of the new fighter and his report was most positive. "The aeroplane is simple and easy to fly and has no vices." On the strength of this, within an unprecedented three months, the Air Ministry gave Supermarine a £1.25 million order for 310 Spitfires, most of which would be used for coastal defence. It was by far the largest order for aircraft that Supermarine had ever received and the firm was too small to handle it. The only option was to sub-contract much of the work in order to get the plane into full production. They began the programme by building only fuselages, while outside firms constructed the other components. Final assembly was done in a Supermarine factory at Eastleigh where the completed aircraft were then flight tested. The assembly process was at times hampered by fitting problems in the early days of subcontracting. In time the Nuffield organisation opened a massive "shadow" factory at Castle Bromwich near Birmingham for the manufacture of Spitfires. It would ultimately produce Lancaster bombers as well, reaching a peak output of 320 Spitfires and 30 Lancasters a month. By June 1940, the Castle Bromwich facility was under the direct management of the Vickers-Armstrong group and had rolled out its first Spitfire, a Mk II. Other aspects of Spitfire manufacture had been dispersed to many locations around southeast England, largely in an effort to protect its production from the attentions of German Air Force bombers.

The Spitfire was the only Allied fighter to remain in both full production and front-line operational service throughout the Second World War.

By September 1937, the Mk I incorporated the Rolls-Royce Merlin F engine producing 1,045 hp. The fighter had been fitted with a Barr and Stroud Type GD5 reflector gunsight and a G22 camera gun. Delivery of the first Mk Is, to No 19 Squadron at Duxford, began on August 6th 1938. That initial Air Ministry order for 310 Spitfires was followed by orders for three more batches totalling an additional 950 aircraft. By the outbreak of the war on September 3rd 1939, nine RAF squadrons had been equipped with Spitfires.

Reg Mitchell knew that if he managed to survive for four years without the cancer recurring, he stood a good chance of living to a normal old age, but by the summer of 1936 he was again suffering considerable pain. In February of 1937 he was examined and tested in a London hospital in the hope that a second operation might help. There would be no second operation. He was told that he had only four or five months to live.

Dr Gordon Mitchell: "After twenty years of happy married life, Mitchell and his wife, Flo, faced the separation which now seemed inevitable. The

British doctors had said there was no hope, and Mitchell had arranged his affairs, leaving no loose ends for those left behind to clear up. During his long illness his wife had always been close beside him, visiting him in hospital and sharing the weeks of convalescence at the seaside after his operation in 1933. She faced the ordeal bravely, and endeavoured to hide her own sorrow as she tried to match her courage with his. But it was hard just to sit and wait, without being able to help.

"March was moving into April, and already the spring flowers were blooming in the 'Hazeldene' garden, when Mrs Mitchell turned to her husband one day and said: 'There must be something more we can do.' " He then told her about a man in Vienna he had heard about who was said to be one of the world's leading cancer specialists. They agreed to go and see the man at his Austrian clinic. Before leaving for Vienna, Mitchell went to the Supermarine works for the last time. He looked through the windows of his office at his view of the River Itchen and then spoke to each member of his design team to thank them for all the good work they had done for him, and to say goodbye. He died at noon on June 11th 1937.

One of the first people that R.J. Mitchell appointed to his design staff at Supermarine was Joseph Smith, a draughtsman who had apprenticed with the Austin automobile firm at Birmingham. Smith would succeed Mitchell as Chief Designer for

Supermarine, a position he would hold until the mid-1950s when he too died of cancer. Dr Gordon Mitchell: ". . . Smith, a man of great energy, determination and courage, fully recognised that he had inherited an aircraft of exceptional quality, and his task would be to develop it to its maximum potential. In the years to come he was to direct the unprecedented development of the Spitfire which was to keep it in the front rank of fighter performance throughout the war years and extend its capabilities over a range of military roles which even Mitchell could not have foreseen." Of Smith, J.S. Scott, author of *Vickers: A History*, wrote: "Although he had been a great admirer of Mitchell, Smith had never tried to imitate his visionary boldness, for his own talent lay in developing things which were already known to be good. If Mitchell was born to design the Spitfire, Joe Smith was born to defend and develop it."

In addition to the Spitfire and the other fine aircraft designs that R.J. Mitchell produced, his legacy to British aviation and the aviation world was the superb design team he had assembled, trained and led. Joe Smith who, like Mitchell, worked a six-and-a-half-day week, continued the pattern of excellence established by R.J., and led the design group into the jet age.

A MK V OF THE HISTORIC AIRCRAFT COLLECTION LTD.

EARLY
SPITFIRE

According to Jeffrey Quill, Chief Production Test Pilot for the Spitfire programme at Supermarine, who was present on the afternoon of March 6th 1936 (not the 5th, as most aviation historians have written), the prototype Spitfire took off on its maiden flight from Eastleigh aerodrome near Southampton, England with Joseph "Mutt" Summers at the controls. Fifteen minutes later the aeroplane was safely back on the ground and there was elation among the small group of official company observers. Quill has substantiated his contention that March 6th was the actual day of the first flight, pointing out that both Major H.J. Payn and Stuart Scott-Hall appear in the photo on page 48 which was taken immediately after K5054's maiden flight, with Summers, R.J. Mitchell, and Quill himself, who had taken both Scott-Hall and Payn up for a brief joy ride in the company's new Miles Falcon that day. He has also affirmed that he flew Mutt Summers to Eastleigh on the 6th specifically for the purpose of Summers making the maiden flight of the prototype on that day.

Quill recalled the scene, noting that K5054 was in its "works finish", unpainted except for its priming coats. It had been fitted with a special fine-pitch propeller for safety in the take-off and to minimise swing due to prop torque during the take-off roll. The aeroplane was well under its maximum take-off weight, with no guns or ammunition installed. There was only a light wind blowing across the airfield and Summers taxied out towards one of the four large Chance lights then set around the field perimeter, turned into the wind, opened the throttle and, after a very

short run, climbed away comfortably. It was a short flight which Summers devoted entirely to checking the slow-flying and stalling characteristics of the aeroplane before bringing it back to land. Following normal procedure on initial flights of aircraft, he left the undercarriage in the down and locked position for the whole flight. "Without too much float," Summers managed a good three-point landing and taxied the prototype to the hangar where the little group of company onlookers waited.

Jeffrey Quill's first opportunity to fly the prototype Spitfire came on March 26th. Ken Scales, who Quill referred to as "nanny" to the prototype, was there on the port wing to help Jeffrey fasten his parachute and Sutton harness. Quill: "The cockpit was narrow but not cramped. I sat in a natural and comfortable attitude, the rudder pedals were adjustable, the throttle and mixture controls were placed comfortably for the left hand, the seat was easily adjustable up or down. The retractable undercarriage selector lever and hydraulic hand pump were situated to the right of the seat. The instrument panel was tidy, symmetric and logically laid out. The windscreen was of curved perspex which gave a good deal of optical distortion but it had a clear view glass panel (not yet armoured) for vision dead ahead in the line of the gunsight. The sliding canopy was straight sided and operated directly by hand with a latch which engaged the top of the windscreen. With the seat in the fully up position there was very little headroom, but at once I felt good in that cockpit.

"I primed the Merlin carefully and

it started first time. I began taxiing out to the north-east end of the airfield which, of course, was entirely of grass. Never before had I flown a fighter with such a very long nose; with the aircraft in its ground attitude vision directly ahead was completely obscured so I taxied slowly on a zigzag course in order to ensure a clear path ahead. The great two-bladed wooden propeller, by this time of maximum coarse pitch, seemed to turn over very slowly and from the stub exhausts, one for each of the twelve cylinders, came a good powerful crackle whenever a small burst of power was applied for taxiing followed by a lot of popping in the exhausts as the throttle was closed again. On arrival at the edge of the field I turned the aircraft 45° off the wind and did my cockpit checks which, at that stage, really consisted only of fuel cocks, trimmer and flap settings, radiator shutter, tightening the throttle friction grip and a quick check over the engine instruments. With a last look round for other aircraft I turned into wind and opened the throttle.

"With that big fixed-pitch propeller able to provide only very low revs during take-off, the acceleration was sluggish and full right rudder was required to hold the aeroplane straight. The torque reaction tended to roll the aircraft on its narrow undercarriage but soon we were airborne and climbing away. At once it was necessary to reset the rudder trimmer and then to deal with the undercarriage retraction and the canopy. This presented a minor problem insofar as the undercarriage had to be raised with a hydraulic hand pump, so it was necessary to transfer the left hand

from the throttle to the stick and operate the hand pump with the right. This was difficult to do without inducing a longitudinal oscillation of the whole aircraft.

"However, once fully airborne and 'tidied up' the aircraft began to slip along as if on skates with the speed mounting up steadily, and an immediate impression of effortless performance was accentuated by the low revs of the propeller at that low altitude. The aeroplane just seemed to chunter along at an outstandingly higher cruising speed than I had ever experienced before, with the engine running over very easily, and in this respect it was somewhat reminiscent of my old Bentley cruising in top gear. I climbed up to a few thousand feet and carried out some steep turns and some gentle rolls and found the aeroplane light and lively but with a tendency to shear about a bit directionally. I put it into a gentle dive and it accelerated with effortless ease and then I came back to rejoin the circuit for landing. The flaps, which I had already tried out in the air, came down on the prototype to only 60° which was the maximum lift, but not maximum drag, position so the glide angle on the approach was very flat and the attitude markedly 'nose up'. This feature was accentuated by the fact that the big wooden propeller ticked over extremely slowly and produced no noticeable drag or deceleration. The approach, with the use of a little power and very 'nose up', meant the view straight ahead was almost non-existent as one got close to the ground, so I approached the airfield in a gentle left-hand turn, canopy open, and head tilted to look round the left-hand side of the windscreen.

Mutt had warned me about this so I was able to get myself on the right line at the outset. As I chopped the throttle on passing over the boundary hedge the deceleration was hardly discernible and the aeroplane showed no desire to touch down—it evidently enjoyed flying—but finally it settled gently on three points, and it wasn't until after the touch-down that the mild aerodynamic buffeting associated with the stalling of the wing became apparent. 'Here', I thought to myself, 'is a real lady.' "

The following day Quill made two flights in the prototype to accurately determine the level top speed at altitudes above and below its optimum full throttle height of 17,000 feet. The aeroplane delivered extremely disappointing results, topping out at 335 mph. R.J. Mitchell was particularly worried as he was aware of the performance of the other new British fighter, the Hawker Hurricane, which had entered flight testing a few months earlier. Mitchell knew that unless the Spitfire could deliver a significantly better performance with a higher top speed than the Hurricane, the Air Staff would not be able to justify a production order for the Spitfire. In his determination to achieve a high level of performance Mitchell had compromised certain important aspects of the Spitfire design, including the pilot's view from the cockpit, the thickness/chord ratio of the wing, and the relative ease of production. Before he would release the aeroplane to Martlesham Heath for the all-important performance trials, he had to find a way to raise its top speed to something well in excess of the Hurricane. His goal was 350 mph. A number of ideas were tried to reduce drag and improve the aerodynamic efficiency, without substantial improvement. Then the Supermarine designers concluded that the fixed-pitch wooden propeller might be the culprit by being subject to compressibility. A new propeller was designed and Quill tested it on the aeroplane on May 5th achieving 348 mph, close enough to 350 for Mitchell to send the prototype off for its date at Martlesham Heath, near Ipswich on the Suffolk coast. Mutt Summers flew the aeroplane up to Martlesham on May 26th where it was put through its paces by three RAF pilots who established that it had a true top speed of 349 mph, a service ceiling of 35,400 feet and a take-off run of 235 yards in a five-knot wind. What the Air Ministry wanted to know was simple: was the Spitfire capable of being flown safely by ordinary squadron pilots? The answer was "yes" and Supermarine received an initial order for 310 aircraft on June 3rd. On that same day Hawker was awarded an initial contract for 600 Hurricanes. But these were very early days for both aircraft. While they both performed well, they were still a long way from being developed into the deadly fighter weapons they would become.

One of Jeff Quill's fondest memories was of the brief private conversations he had with R.J. Mitchell on the airfield at Eastleigh. After most of his early test flights in the prototype Spitfire, Quill would be met at the aeroplane by Mitchell who had given strict orders that he was to be informed by telephone whenever the plane was about to fly so that he

could be present. Each time, as Quill entered the Eastleigh landing circuit he would scan the field to see if Mitchell's yellow Rolls-Royce (a gift from R-R after the British victory in the 1931 Schneider Trophy race) was present on the tarmac. If no significant problems had arisen during the test flight Mitchell would normally invite Quill to sit and talk with him in the Rolls while they waited for the aeroplane to be made ready for another flight. Quill was extremely conscious of his own lack of training in aeronautical engineering and, though a superbly trained and experienced pilot, he was ever aware of the exceptionally high level of technical expertise among the Supermarine design staff. Mitchell greatly valued Quill's contribution to the Spitfire programme and often used the sessions in the car to bolster the pilot's morale and confidence. Mitchell: "Jeff, if anyone tries to tell you something about an aeroplane which is so damn complicated that you can't understand it you can take it from me it's all balls."

Mitchell's cancer returned in 1936 and his health declined rapidly. In February 1937 he decided to try treatment at the Anton Loew Clinic in Vienna. Quill: "R.J. returned to England late in May with any last hope of recovery finally gone. There was nothing for it but to await the end, which he did with the utmost fortitude. I drove from Brooklands to Southampton shortly after his return to see him. He was lying on a day-bed in the sitting room of his house in Russell Place and he talked animatedly and cheerfully, but when I left I knew I would not see him again. When he died on 11th June, at the

PREVIOUS SPREAD: AR213 SHOWING
ITS UNDERSIDE OVER BUCKING-
HAMSHIRE; ABOVE: THE NEWLY
RESTORED MK 1A ON A TEST FLIGHT
OVER OXFORDSHIRE.

age of only forty-two, no production Spitfire had then taken the air, but he had seen the prototype of his fighter fly many, many times and knew that it was doing what he had hoped and that it was ordered into quantity production. He had every reason to expect that it would serve the Royal Air Force and his country well. It is appropriate to recall that it was during those three years following his major operation in 1933 that R.J. designed the Spitfire, knowing that

his life might be cut short at any time. He was a brave man as well as a brilliant designer."

Building the Spitfire in production quantities proved to be a far more complex and demanding effort than anyone had anticipated. The techniques and methods of the manufacturing and large-scale assembly process were not part of the company's experience, nor of most aircraft makers of the time, and the Air Council had lit-

tle patience with Supermarine when it fell behind schedule in the delivery of aircraft from the initial order of 310 planes. By late April 1939, the company was fully six months behind its forecasted delivery date for the 150th aircraft of the order. Relations between the Air Council and Supermarine's Sir Robert McLean deteriorated in this time as some Council members came to believe that McLean was unnecessarily contributing to the Spitfire delays by his deter-

mination to keep the production work inside the company rather than sourcing enough of the work out to subcontractors. As it happened, their view was not justified, but the Council was under pressure to bring the Spitfire to operational status in the greatest numbers possible. The British Foreign Office intelligence sources were now advising that the German Air Force might well be capable of going to war as early as January 1939 as opposed to their earlier estimate of

1942. And it was not only Spitfire production that was lagging; delivery of the Hurricane, with an initial order of twice the number of aircraft, was also well behind schedule. The Hurricane, however, was much easier to build, having been designed (by Sydney Camm) for relative simplicity of production, and in a short time the Hurricane production caught up with the delivery schedule.

In fact, the British government had been as much a cause of the delays as the manufacturers. Its policy since the end of the First World War had been to maintain a substantial number of firms in the aircraft industry in peace-time, providing small orders to keep them in business and therefore available for expansion in any new national emergency. Good in theory, the policy had a downside as well. When that emergency arose later in the 1930s, none of the plane makers was experienced in large-scale production and were simply unable to respond efficiently and dramatically "overnight" as the Air Council expected.

While Supermarine did retain the construction of Spitfire fuselages and the final assembly and flight testing operations, the wings and tail units were sourced from subcontractors. The early delays were caused by late deliveries of the wings to Supermarine. The wing subcontractors blamed their lateness on poor-quality drawings supplied by Supermarine. Just as it was true that Britain's aircraft industry was not yet experienced in the most modern manufacturing techniques, the same was true of the subcontractors to that industry. No one was really prepared to meet the requirements for full Spitfire production in June 1936 when the entire Supermarine work-

force amounted to no more than 1,370 people. Supermarine would grow and change and become part of Vickers-Armstrong. New management and leadership emerged to resolve the difficulties of bringing Spitfire production up to the acceptable quality and delivery standard. A new general manager, H.B. Pratt, came over from the Vickers parent organisation and it became his task to solve the Spitfire production problems. Under his supervision Spitfire production was back on schedule by March 1939.

Jeff Quill had a hand in the improvements when he and fellow Supermarine test pilot George Pickering co-developed a flying test schedule for Spitfire production which, with modification, lasted for ten years through the many marks of the type. Quill: "It was based on the principle that, in addition to the essential functional tests of the whole aeroplane (which were listed), the basic performance and handling should also be checked at selected points throughout the flight envelope. This meant that each aeroplane was to be climbed at maximum continuous climbing power and best climbing speed to its full throttle height of at least 18,000 feet and all instrument readings checked. This was to be followed by a two-minute level run at maximum combat power settings to check the indicated top speed and the performance of engine and supercharger. The aeroplane was then to be put into a full-power dive to its limiting indicated airspeed of 470 mph and its trim and control behaviour checked in this extreme condition. In general, and allowing for two or three short initial flights for trimming the ailerons and adjusting engine

boost and propeller settings, this schedule took about 40 minutes' flying time and gave the aeroplane a thorough shakedown. A minimum of three take-offs and landings was also required before the aeroplane could be passed for delivery. There would be no concessions on any aspect of the schedule."

R.J. Mitchell's long-time colleague in the Supermarine design department was Joe Smith, who had contributed enormously to the design of the Spitfire and would succeed Mitchell as Chief Designer. Perhaps no one had more faith in the aeroplane or reason to believe in the capability of its design, which would be extensively developed and evolved over its long production history. Smith's vision, strength of personality, his great design skill and his temperament took the basic design wherever it needed to go and led to his promotion to Manager of the Design Department. J.D. Scott wrote of Smith, "If Mitchell was born to design the Spitfire, Joe Smith was born to defend and develop it." Jeffrey Quill has explained the use of the word defend in that quote. "Although much liked by pilots from the outset the Spitfire never found much real favour with the Air Council until it had decisively proved its mettle in battle over Dunkirk. Originally many technical people were suspicious of it, many production advisers in the Air Ministry did not care for it, and the Air Council were outraged by the delays in production during the latter part of 1937 and during 1938.

"On 7 June 1939 a memorandum was sent to the Chief of the Air Staff by the Air Member for Development and Production, Sir Wilfrid Freeman, in which he referred to 'orders to be placed now with certain firms whose existing orders will run out early in 1940'. And on the subject of Supermarine he wrote: 'Supermarine will run out of their order for Spitfires in February or March 1940 and since it will be impossible to get a new aircraft into production at Supermarine before September 1940 there is certain to be a six-month gap which we will have to fill.

" 'In order to be able to bridge the gap with as few machines as possible, Supermarine will be told later on to reduce the amount of subcontracting and get their men onto single shift so that although Supermarine production is likely towards the end of the present contract to exceed 48 aircraft per month it is hoped that we can reduce the gap production to 30 aircraft a month. Vickers are pressing for a more generous release of Spitfires for foreign orders, and it seems to me that provided no releases are made until October, we could go some way to meet them this year and could release aircraft for foreign orders freely after the spring of next year, when the Castle Bromwich factory will be coming into production. The type of aircraft that could be put into production at Supermarine after the end of their contract would be Beaufighter, Gloster Fighter, Lysander or Westland (F.37/35).'

"Two things are clear from this memorandum. First, that although the Mk I Spitfire had been in squadron service for nearly a year the Air Staff had really failed to appreciate the potential of the aircraft they had in their hand. Second, by planning to allow Spitfire production to phase out at Supermarine (in favour of the

Beaufighter) they obviously had no ideas whatever for developing the Spitfire beyond the Mk I stage, nor had they much appreciation of the way operations would develop when hostilities broke out. The whole tone of this memorandum suggests that the Air Council had already decided they were not much interested in going on with the Spitfire beyond their current contractual obligation to Supermarine, and that the planned shadow factory at Castle Bromwich could look after any additional production that might be required so the only outstanding question was what to do with Supermarine. Castle Bromwich had been planned to produce bombers as well as fighters so the Air Council had flexibility of choice at that time.

"It is a sobering thought that this talk of phasing out the Spitfire was going on only 14 months before the start of the Battle of Britain! That was what Scott meant when he wrote of Joe Smith having to defend as well as develop the Spitfire."

The first production Spitfire on the line at Castle Bromwich was the Mk II, essentially a lightly refined Mk I engined by a slightly more powerful Merlin and fitted with a new constant-speed propeller with processed wood blades made by the new firm Rotol, a creation of Rolls-Royce and Bristol Aviation formed to increase the production of propellers in Britain during the war. In the area of aero engine development, Rolls was busy reviving developmental work on its "R"-type engine that had powered the Supermarine S.6 Schneider Trophy-winning seaplane. The military potential of the engine was too great to ignore and, while continuing

to develop the Merlin to ever more powerful levels, R-R focused, too, on the future of the larger R-type powerplant which they called the Griffon. Even before the war had broken out in 1939, Joe Smith was excited about the prospects of the Griffon and considering how to marry the engine to the Spitfire.

The gigantic new shadow factory at Castle Bromwich was initially operated by the Nuffield organisation whose main business was Morris Motors. The Nuffield people quickly discovered that building state-of-the-art fighter planes in wartime was not much like the mass production of automobiles in peacetime and serious troubles soon arose in the Birmingham facility. The situation was reaching a crisis point when Lord Beaverbrook, the new Minister for Aircraft Production in the Churchill government, decided to turn the management of the Castle Bromwich factory over to the Vickers-Armstrong management who immediately began the coordination of Spitfire production between Supermarine in Southampton and Castle Bromwich. The pace of production at Castle Bromwich picked up impressively and would eventually reach a massive output of 320 Spitfires and 30 Lancasters a month.

Jeff Quill and his fellow test pilots' workload at Supermarine was increasing significantly and they needed additional help. To meet his standards, Quill insisted on making the choice of a new pilot himself. Alex Henshaw had by this time earned a glowing reputation via his three record flights between London and Capetown in an aircraft roughly half the size of a Spitfire. Quill was

impressed by Henshaw's achievements and thought the record-breaker might be just the pilot he was looking for. They met and got on well from the start. Henshaw was offered the position by Quill and accepted it, arriving at Eastleigh a few days later. Soon, as Jeff watched Henshaw's amazing flying skills he realised that in Alex he had the man to run the test flying programme at Castle Bromwich.

"There is no doubt that the job of Chief Pilot at Castle Bromwich was one of exceptional difficulty and challenge. It was fortunate for us all that Alex was there to do it. I know of no other pilot who could have handled it as he did, nor who could have set such a standard for his subordinate pilots to follow. It was a case of leadership by example if ever there was one. I refer, of course, to the formidable task of flight testing such a huge output of aircraft in abominable weather with no 'aids.'

ANOTHER FINE VIEW OF THE MK 1A AR213 ON A POST-RESTORATION TEST FLIGHT.

SUMMER 1940

It would be remembered as an exceptional period of gloriously hot days and brilliant azure skies, that uncharacteristically warm English summer of 1940. It would be remembered, too, for the Battle of Britain, the most spectacular air campaign of all time. Four years earlier German Chancellor Adolf Hitler's jackbooted troops had occupied the Rhineland. His Condor Legion airmen had gone to the aid of General Francisco Franco in the Spanish Civil War where they tried out new weaponry, including a Messerschmitt fighter plane that would be seeing a lot of action in the coming years. In March 1938, three days after the Nazis annexed Austria, Winston Churchill had spoken at length in the British House of Commons on the urgency of resisting the German leader. On September 15th British Prime Minister Neville Chamberlain emerged from a meeting with the German Chancellor, declaring that Herr Hitler "appeared to be a man who could be relied upon when he had given his word". At the end of the month The Munich Agreement allowing Germany to annex the Sudetenland portion of Czechoslovakia was signed by Hitler, Chamberlain, French Premier Edouard Daladier and Italian Premier Benito Mussolini. While in Munich, Chamberlain and Hitler also signed a document pledging that the peoples of Germany and Britain would never go to war with one another again. They further resolved to consult with each other on matters of mutual concern and to contribute to peace in Europe. The following day Chamberlain returned to London. At Heston airfield, near what is now Heathrow airport, he waved the document before the assembled press and told them: "This is the second time there has come back from Germany to Downing Street peace with honour. I believe it is peace for our time."

Hitler, of course, invaded Poland on September 1st 1939; Britain and France declared war on Germany on the 3rd and that same day Churchill joined Britain's War Cabinet as First Lord of the Admiralty. On March 2nd 1940, the first naval action in the English Channel occurred when bombers of the German Air Force attacked the passenger vessel SS *Domala*. Less than two weeks later the first British civilian to be killed in the conflict died during a German air raid at Scapa Flow. In April, the Nazis invaded Denmark and Norway and on May 10th they took the Netherlands. That same day Neville Chamberlain resigned as Prime Minister and was replaced by Winston Churchill.

During that same week in May,

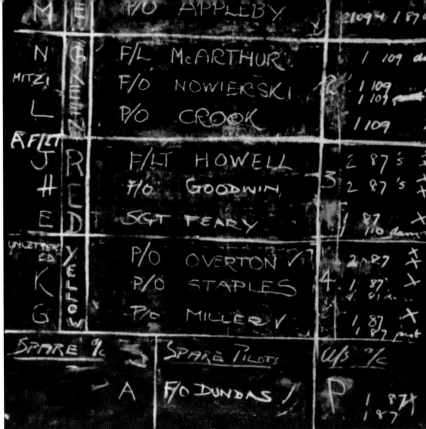

German forces invaded Belgium and France. Hurricane fighters of the Royal Air Force put up fierce resistance from their French airfields, but were gradually overwhelmed and on May 22nd had to abandon their last French base and return to England. The German push towards Dunkirk on the Channel coast drove the troops of the British Expeditionary Force onto the beaches where they were trapped while awaiting evacuation which began on May 26th. The rescue operation was successful, safely returning nearly all of the British soldiers to the port of Dover. In two more days Belgium surrendered to the Germans. The Battle of France began and raged until June 22nd when France fell to the Nazis. It was the turn of Britain to face the full might of Hitler's air power. In the House of Commons Churchill noted that wars were not won by evacuations and forecast that the cause of civilisa-

tion itself would be defended by the skill and devotion of a few thousand airmen. As Group Captain Peter Townsend later commented "The hour, the finest hour, of the whole British people had struck. It had struck too, more urgently, for their first-line of defence, Fighter Command, as yet untried in the role." Townsend also noted that the Battle of Britain was a victory for the whole British people—the man and woman and child in the street, the civil defence units, the Navy, the Army, and the Bomber and Coastal Commands of the RAF "We 'few' happened to possess the necessary weapons to fight the enemy hand to hand. Those weapons were the Spitfire and the Hurricane."

In his monumental account, *The Second World War*, Churchill wrote: "The buoyant and imperturbable temper of

Britain, which I had the honour to express, may well have turned the scale. Here was this people, who in the years before the war had gone to the extreme bounds of pacifism and improvidence, who had indulged in the sport of party politics, and who, though so weakly armed, had advanced lightheartedly into the centre of European affairs, now confronted with the reckoning alike of their virtuous impulses and neglectful arrangements. They were not even dismayed. They defied the conquerors of Europe. They seemed willing to have

FAR LEFT: THE CREW OF A HEINKEL HE-111 BOMBER DURING A RAID ON AN ENGLISH TARGET: CENTRE: A PRESENTATION PLATE HONOURING KG-55, A GERMAN BOMBER SQUADRON IN 1940; RIGHT: A STATE BOARD OF 609 SQUADRON DURING THE BATTLE OF BRITAIN.

THE STORY OF THE PLANE THAT BUSTED THE BLITZ!

SPITFIRE

their island reduced to a shambles rather than give in. This would make a fine page in history."

As France neared capitulation in June, the new British Prime Minister recognized the danger of allowing his last military reserves, including his fighter squadrons, to be sacrificed in aid of the futile French resistance. Yet such was the political and emotional capital that he continued to invest in his cross-Channel ally, repeatedly urging her to fight on, that he felt power-

fully obliged to assist her. At the same time he felt constrained to keep back what he believed to be a sufficient fighter reserve of at least twenty-five squadrons for the defence of England. In the end, it was mainly due to the determination of the Air Officer Commanding-in-Chief, Fighter Command, Air Chief Marshal Sir Hugh Dowding, that the minimal number of fighter squadrons were retained on English airfields for operations in the Battle of Britain.

The Battle, the first major confrontation to be fought solely in the air, was made necessary by Hitler's late-blooming decision to invade Britain. For this to succeed, he required total air superiority over the Royal Air Force, a task he assigned to the head of his air force, Reichsmarschall Hermann Göring, who had flown Fokker D-VII fighters in World War One. As an early and ardent Nazi, Göring had risen to power alongside Hitler. He assured his Führer that British air power and coastal defences would be quickly destroyed.

July 10th 1940 is generally accepted as the day the Battle began. For all of the preceding nine months, which had included one of the coldest winters of the century, the pilots of RAF Fighter Command waited through what was called the Phoney War or Sitzkrieg for the combat they knew was coming. Late in September 1939 the British Army, under the command of Lord Gort, had crossed the English Channel to France accompanied by four squadrons of RAF Hurricanes, eventually backed up by two squadrons of Gloster Gladiator fighters. No Spitfires were sent to France in the period prior to the Battle of Britain. Six days after the opening round of the Battle, Adolf Hitler issued his Directive 16 outlining his plan for the German invasion of Britain, Operation Sea Lion.

Adolph 'Sailor' Malan was a great fighter pilot and probably the greatest RAF fighter leader of the Second World War. "He was a born leader and natural pilot of the first order. Complete absence of balderdash. As far as he was concerned, you either did your job properly, or you were on your way. He inspired his air crews by his dynamic and forceful personality, and by the fact that he set such a high standard in his flying."
– Pilot Officer W.M. Skinner, 74 Squadron

Malan's skill in air gunnery was unmatched by any other pilot in his squadron; possibly in the entire air force. The South African was naturally gifted in the art of deflection shooting. He absolutely 'knew' where his bullets were going and how to lead them into an enemy aircraft. He was also a superb marksman from nearly any reasonable distance, but he invariably pressed his attacks, getting as close to his prey as he could. Of his early flights in the Mk 1 Spitfire that equipped his squadron at Hornchurch, he recalled: "It was like changing over from Noah's Ark to the Queen Mary. The Spitfire had style and was obviously a killer. We knew that from the moment when we first fired our eight guns on a ground target. Moreover, she was a perfect lady. She had no vices. She was beautifully positive. You could dive till your eyes were popping out of your head, but the wings would still be there—till your inside melted, and she would still answer to a touch." The Mk 1 that he flew was much lighter and handled better than any of the later Spitfire marks. It was well-armed and delivered a top speed of 362 mph with a rate of climb of 2,000 feet a minute. Malan: "You fix the [machine-gun] range on the ground. 250 yards is the deadliest. The idea behind this armament is that each gun has its own little place in the heaven. By a criss-cross of fire, ranged at a selected distance, you achieve the maximum lethal pattern."

Sailor's combat career did not begin in glory. On September 6th 1939 his flight was sent on an interception of a suspected German bomber raid. The unidentified "enemy raiders" turned out to be Hurricanes of another RAF fighter squadron on their way back to England after an operation. Tragically, the primitive radar plot of the time led Malan to order his pilots into an attack and two of the Hurricanes were shot down. The 'friendly fire' incident later became known as the 'Battle of Barking Creek'. The two pilots of Malan's flight who downed the Hurricanes were court-martialed. Malan testified that he had ordered a recall prior to the action, which was found to have been an accident.

At the start of the Battle of Britain, the destruction of a Heinkel He 111 bomber was shared by Sailor on July 12th. On the 19th he shot down a Bf109 fighter and damaged two more Bf109s in the following week. A mechanic assigned to Malan's aeroplane recalled: "Having spent many hours patching up his Spitfire ready for the next trip I could well realise the marvelous escapes he must have had. Although his Spitfire came back battered each time, he would not part with it in exchange for a new and more modern one. His instructions to his crew were: 'My machine has got to be serviceable. There is no excuse.' His engine had to go first time, the radio-telephone just had to function even if his junior pilots' radios failed at times. And his guns weren't allowed to have stoppages. On one occasion it was my job to work out in the open all night with a hand torch to renew his battered tail-plane. I don't know quite how I managed it, but I knew it just had to be done by 4 a.m. Flight Lieutenant Malan got in his cockpit and said: 'Contact' without asking if I had finished the job. In fact, I was struggling with the last stubborn split pin. The day came when we were shown the films of his combats, which was a tonic to us all after eight months of terrible waiting, but always ready. The greatest thrill of all was the night of the first raids, when Lt. Malan went up alone through the intense gunfire and shot down two German machines in what seemed less than ten minutes."

LEFT: ATA DELIVERY PILOT MONIQUE AGAZARIAN; OTHER SPITFIRE PILOTS: BELOW: JOHN BURGESS, TOP RIGHT: CARROLL MCCOLPIN, FAR RIGHT: BRIAN LANE AND COLLEAGUES, BOTTOM RIGHT: A. N. MACGREGOR.

In the second month of the Battle, Sailor was made commander of No 74 Squadron. His deputies were H. M. Stephen and J. C. Mungo-Park. He set about the job of developing his pilots into "the toughest bunch in Fighter Command," and his policy with them was "Kick their arses once a day." As the Battle wore on, he drove himself as hard as he drove them. Beginning at 7 a.m. August 11th, his third day in command of the squadron, they flew the first of four operations against enemy raids approaching the English south coast. By the end of the day they had collectively downed thirty-eight German aircraft. The day was thereafter referred to as 'Sailor's August Eleventh.'

His greatness as fighter pilot, tactician and air leader is exemplified by his famous Ten Rules for Air Fighting, basic tenets for the fighter pilots of the Royal Air Force in that time. Pilots who paid attention to Sailor's rules often owed their survival to him.

"Generally speaking, tactics in air fighting are largely a matter of quick action and ordinary commonsense flying. Apart from keeping your eyes open wide and remaining fully alive and awake, it is largely governed by your own aircraft in comparison with that flown by your opponent. In the case of the Spitfire versus the Me 109F, the former has superior maneuverability, whereas the latter has a faster rate of climb. The result is that the Spitfire can afford to 'mix it' when attacking, while the Me 109F, although it tends to retain the initiative because it can remain on top, can-

TOP LEFT: AN RAF RADIO TECHNICIAN; TOP CENTRE: AMERICAN EAGLE SQUADRON PILOT ART ROSCOE; TOP RIGHT: ALAN MITCHELL; BOTTOM LEFT: ATA PILOT DIANA BARNATO-WALKER; BOTTOM RIGHT: COLIN GRAY AND AL DEERE.

MARK HANNA TAXIING OUT IN MH434, THE SPITFIRE OF THE OLD FLYING MACHINE COMPANY, AT DUXFORD, CAMBRIDGESHIRE.

BRIAN KINGCOME

HUGH DUNDAS

not afford to press the attack home for long if the Spitfire goes into a turn. Obviously, there are a lot of factors involved which must govern your action in combat—such as the height at which you are flying, the type of operation on which you are engaged, the size of your formation, etc. There are, however, certain golden rules which should always be observed:

1. Wait until you see the whites of his eyes. Fire short bursts of one to two seconds only when your sights are definitely 'ON'.
2. Whilst shooting think of nothing else, brace the whole of your body;

have both hands on the stick. Concentrate on your ring sight.
3. Always keep a sharp lookout. 'Keep your finger out!'
4. Height gives you the initiative.
5. Always turn and face the attack.
6. Make your decisions promptly. It is better to act quickly even though your tactics are not the best.
7. Never fly straight and level for more than thirty seconds in the combat area.
8. When diving to attack, always leave a portion of your formation above to act as a top guard.
9. INITIATIVE, AGGRESSION, AIR DIS-CIPLINE and TEAM WORK are words

that mean something in Air Fighting.
10. Go in quickly—Punch hard—Get out!

"Taking a Spitfire into the sky in September 1940 was like entering a dark room with a madman waving a knife behind your back."
–Adolph 'Sailor' Malan

One of Churchill's better decisions as Prime Minister was to appoint William Maxwell Aitken, the first Baron Beaverbrook, to the newly-created post of Minister of Aircraft Production. The bumbling and the frequently archaic

JOHN MUNGO-PARK

BRENDAN FINUCANE

attitudes of the Air Ministry had left Britain dangerously under-prepared for the war and the air defence of the country. Beaverbrook worked sixteen hours a day from Stornoway House, his London home on St James Park. Of him, Lord Dowding once said: "I saw my reserves slipping away like sand in an hour glass . . . without his drive behind me I could not have carried on in the Battle." Thanks mainly to Beaverbrook's ruthless approach to his task, the RAF was never short of the fighters it needed to continue the Battle; what it lacked was a sufficient number of pilots to fly them. But

Churchill's scientific adviser, Professor Frederick Lindemann, concluded that by drastically reducing the pilots' operational training time from six months to four weeks, the monthly output of pilots could be increased from 560 to 890. Lindemann: "Are not our standards of training too high? The final polish should be given in the squadrons."

Early in the Battle the pilots of 19 Squadron had taken off from their base at Fowlmere in Cambridgeshire to intercept a flight of Dorniers. Flying Officer Jimmy Coward's was

one of the few cannon-equipped Spitfires then operational. In the initial encounter his cannons jammed and, almost at the same instant, he felt pain "like a kick on the shin in a Rugger scrum." When he looked down he was shocked to see his bare left foot lying on the floor of the cockpit, severed from his damaged leg except for a few ligaments. He soon managed to bale out but with the action came a new agony as his nearly-detached foot spun continuously on the remaining ligaments as he descended slowly from 20,000 feet. He could only watch fearfully as his blood sprayed from his tibial artery. He was unable to retrieve the first aid kit from his breast pocket but somehow raised his left leg nearly to his chin and fashioned a crude tourniquet from the radio-telephone lead in his flying helmet, which he then struggled to tie tightly around his thigh. Within the next hour Coward was in the care of surgeons at the Addenbrookes Hospital, Cambridge, where his left leg was amputated below the knee.

Minutes and seconds counted in the demanding days of the Battle and the ground crews often kept their Spitfires airworthy, through clever improvisation. Flying Officer Robert Lucy, the engineering officer of 54 Squadron at Hornchurch, once removed the armour plating from the seat-back of a badly-damaged Spitfire and persuaded a local garage to fashion it into two heavy-duty fishplates which he then used to patch large holes in the wing root of another Spitfire. Lucy took the precaution of removing the plates each evening to be certain that the fairing was holding up under the strain. The aircraft was still operational near the end of the Battle in October. At Duxford, just down the road from Fowlmere, Leading Aircraftman William Eslick performed a bit of time-saving front-line redesign. He switched the access point of the compressed air bottles powering the machine-guns from a nearly inaccessible trap in the cockpit floor to behind the pilot's seat, making far easier access through the sliding hood. In another example of field ingenuity, Aircraftman Harold Mead quickly repaired an eighteen-inch gash in a Spitfire wing by cutting and shaping a large section of a petrol can which he

then fitted in place with four rivets.

"Having mastered the cockpit drill, I got in and taxied out on the aerodrome, sat there for one moment to check that everything was o.k., and then opened up to full throttle. The effect took my breath away. The engine opened up with a great smooth roar, the Spitfire leapt forward like a bullet and tore madly across the aerodrome, and before I had realized quite what happened I was in the air. I felt as though the machine was completely out of control and running away with me. However, I collected my scattered wits, raised the undercarriage, and put the airscrew into coarse pitch, and then looked round for the aerodrome, which to my astonishment I saw was already miles behind."
—Flight Lieutenant David M. Crook

222 Squadron Spitfire pilot Sergeant John H.B. Burgess recalled: "You got that horrible feeling down in the pit of your stomach . . . and when you were climbing you still had that sort of peculiar tummy feeling. But once action started you were too busy and all you were interested in was avoiding getting killed or trying to shoot down the other aircraft. It was rather like a dare to some degree. You wanted to see how far you could go without . . . coming to any harm. If you got caught and shot at and had to do a forced landing, you lived to fight another day . . . I think that the spirit of the successful fighter pilot was to 'look everywhere' and to never be intimidated by the number of enemy aircraft that were around because you didn't realize at the time that they were more frightened than you were . . .

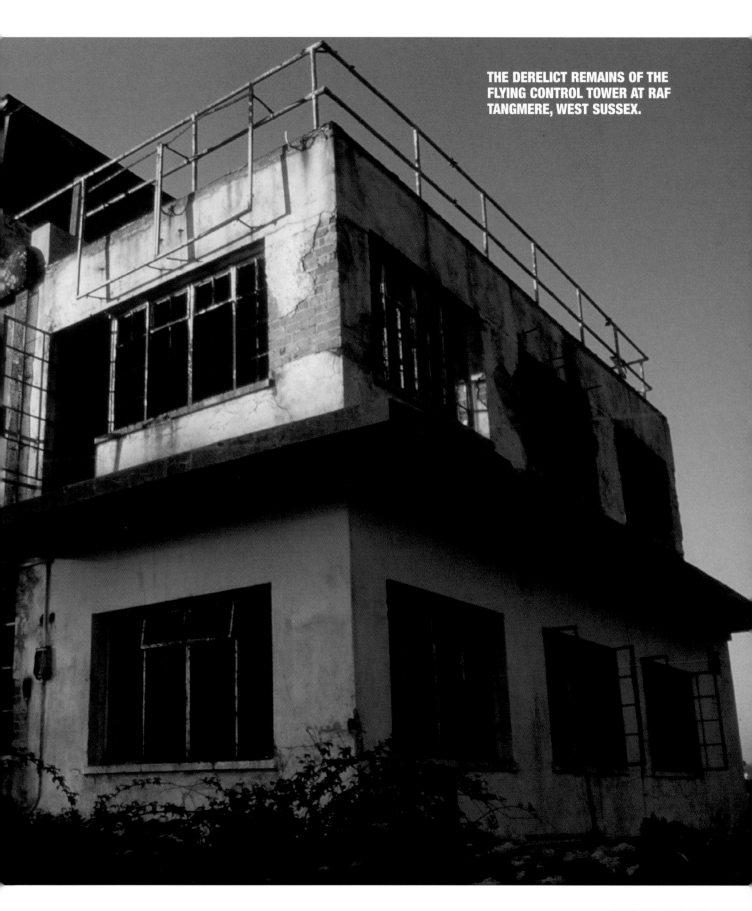

THE DERELICT REMAINS OF THE
FLYING CONTROL TOWER AT RAF
TANGMERE, WEST SUSSEX.

TOWERS OF THE BRITISH CHAIN HOME RADAR NETWORK IN THE BATTLE OF BRITAIN PERIOD; RIGHT: VAPOUR TRAILS TELL OF FRANTIC AERIAL COMBAT OVER LEWES IN THE SYMMER OF 1940.

they were miles away from home, deep into enemy territory . . . if they were caught alone, they were finished." On August 30th, at the height of the Battle, RAF Fighter Command was able to field 372 Spitfires and 709 Hurricanes available for front-line operations. In general, the slightly slower Hurricanes were assigned to attack the enemy bombers, while the Spitfires were sent up to engage their fighter opposition. Both aircraft did their jobs well; the lovely and charismatic Spitfire received most of the richly-deserved glory while the Hurricane accrued the higher score of enemy kills, as would be expected with nearly twice as many Hurricanes in the hunt. The solid, stable Hurricane was the better gun platform; the agile Spitfire the better dogfighter. The pilots of Fighter Command appreciated the virtues of both aircraft, but most loved the Spitfire for the joy it was to fly. However, it was by no means perfect. When damaged through enemy action, the partially fabric-covered Hurricane could often be easily patched up and returned to duty whereas the stressed-metal airframe of the Spitfire required the specialist care of unique servicing units at designated "Spitfire bases." The Spitfire was, however, a better match for the Messerschmitt 109 and, as such, was assigned that challenging role.

Flying Officer Richard Hope Hillary flew Spitfires as a member of No 603 Squadron at Hornchurch in the Battle of Britain. After destroying a Bf 109 fighter on September 3rd 1940, Hillary was himself shot down in flames off Margate. He baled out and was rescued by the Margate lifeboat.

Terribly burned, he became a patient of the pioneering plastic surgeon Archibald McIndoe at the Queen Victoria Cottage Hospital, East Grinstead, where he became one of McIndoe's original 'Guinea Pigs.' He wrote one of the finest books of the war, *The Last Enemy*, published in June 1942. Returning to operations after converting to night fighters, he was killed on January 8th 1943 when the Blenheim he was flying spun into the ground. Hillary: ". . . we learned, finally, to fly the Spitfire. I faced the prospect with some trepidation. Here for the first time was a machine in which there was no chance of making a dual circuit as a preliminary. I must solo right off, and in the fastest machine in the world.

"One of the squadron pilots took me up for a couple of trips in a Miles Master, the British trainer most similar to a Spitfire in characteristics.

"I was put through half an hour's instrument flying under the hood in a Harvard, and then I was ready. At least I hoped I was ready. Kilmartin, a slight dark-haired Irishman in charge of our Flight, said: 'Get your parachute and climb in. I'll show you the cockpit before you go off.'

"He sauntered over to the machine, and I found myself memorizing every detail of his appearance with the clearness of a condemned man on his way to the scaffold—the chin sunk into the folds of a polo sweater, the leather pads on the elbows, and the string-darned hole in the seat of his pants. He caught my look of anxiety and grinned.

" 'Don't worry; you'll be surprised how easy she is to handle.'

"I hoped so.

"The Spitfires stood in two lines outside 'A' Flight pilots' room. The dull grey-brown of the camouflage could not conceal the clear-cut beauty, the wicked simplicity of their lines. I hooked up my parachute and climbed awkwardly into the low cockpit. I noticed how small was my field of vision. Kilmartin swung himself onto a wing and started to run through the instruments. I was conscious of his voice, but heard nothing of what he said. I was to fly a Spitfire. It was what I had most wanted through all the long dreary months of training. If I could fly a Spitfire, it would be worth it. Well, I was about to achieve my ambition and felt nothing. I was numb, neither exhilarated nor scared. I noticed the white enamel undercarriage handle. 'Like a lavatory plug,' I thought.

" 'What did you say?'

"Kilmartin was looking at me and I realized I had spoken aloud. I pulled myself together.

" 'Have you got all that?' " he asked.

" 'Yes, sir.'

" 'Well, off you go then. About four circuits and bumps. Good luck!'

"He climbed down.

"I taxied slowly across the field, remembering suddenly what I had been told: that the Spitfire's prop was long and that it was therefore inadvisable to push the stick too far forward when taking off; that the Spitfire was not a Lysander and that any hard application of the brake when landing would result in a somersault and immediate transfer to a [Fairey] 'Battle' squadron. Because of the Battle's lack of power and small armament this was regarded by everyone as the ultimate disgrace.

"I ran quickly through my cockpit drill, swung the nose into wind, and took off. I had been flying automatically for several minutes before it dawned on me that I was actually in the air, undercarriage retracted and half-way round the circuit without incident. I turned into wind and hauled up on my seat, at the same time pushing back the hood. I came in low, cut the engine just over the boundary hedge, and floated down on all three points. I took off again. Three more times I came round for a perfect landing. It was too easy. I waited across wind for a minute and watched with satisfaction several machines bounce badly as they came in. Then I taxied rapidly back to the hangars and climbed out nonchalantly. Noel [Agazarian], who had not yet soloed, met me.

" 'How was it?'

"I made a circle of approval with my thumb and forefinger.

" 'Money for old rope,' I said.

"I didn't make another good landing for a week.

"The flight immediately following our first solo was an hour's aerobatics. I climbed up to 12,000 feet before attempting even a slow roll. Kilmartin had said 'See if you can make her talk.' That meant the whole bag of tricks, and I wanted ample room for mistakes and possible blacking-out. With one or two very sharp movements on the stick I blacked myself out for a few seconds, but the machine was sweeter to handle than any other that I had flown. I put it through every manoeuvre that I knew of and it responded beautifully. I ended with two flick rolls and turned back for home. I was filled with a sudden exhilarating confidence. I could fly a Spitfire; in any position I was its master. It remained to be seen whether I could fight in one.

"It had happened. My first emotion was one of satisfaction, satisfaction at a job adequately done, at the final logical conclusion of months of specialized training. And then I had a feeling of the essential rightness of it all. He was dead and I was alive; it could so easily have been the other way round; and that would somehow have been right too. I realized in that moment just how lucky a fighter pilot is. He has none of the personalized emotions of the soldier, handed a rifle and bayonet and told to charge. He does not even have to share the dangerous emotions of the bomber pilot who night after night must experience that childhood longing for smashing things. The fighter pilot's emotions are those of the duelist—cool, precise, impersonal. He is privileged to kill well. For if one must either kill or be killed, as now one must, it should, I feel, be done with dignity. Death should be given the setting it deserves; it should never be a pettiness; and for the fighter pilot it never can be."

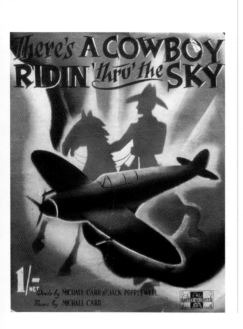

SPITFIRE PILOT
RICHARD HILLARY,
AUTHOR OF THE
LAST ENEMY, ONE OF
THE FINIST BOOKS
OF THE WAR; RIGHT:
VARIOUS BATTLE OF
BRITAIN FIGURES.

GEOFFREY PAGE

ADOLF GALLAND

HEINZ BAER

HERMANN GOERING

ROBERT STANFORD TUCK

HUGO SPERRLE

ROBERT OXSPRING

HUGH DOWDING

GUNTHER RALL

"The Spitfire flew as it had looked on the ground—a sheer dream; the controls beautifully harmonised, positive and quick response; like a true thoroughbred, not a vice in the whole machine. I felt part of it and knew at once that I would give everything I possessed to make the combination successful."
– Alex Henshaw, Chief Test Pilot, Castle Bromwich

"We continued flying our mixed bag of aircraft until about the twentieth of March, when a solitary Spitfire landed and taxied over to our hangar. We were told it was ours. Our hearts leapt! We walked around it, sat in it, and stroked it. It was so beautiful I think we all fell a little bit in love with it.

"In the next few days, a further fifteen Spitfires arrived and for the first time we were a real squadron. All we had to do was to fly them. About that time a message arrived stating that anyone crashing a Spitfire would be posted immediately. We thought this a little hard in view of the numbers we had seen crashed on the airfield during the previous few months."
– Wing Commander Bob Doe

No one has flown or is more familiar with the Mk Ia Spitfire AR213 than

FLYING SPITFIRE

aerobatic and warbird pilot Tony Bianchi. Tony has been tasked by the various owners of the aeroplane to maintain and fly it almost continuously since the mid '70s. "I have about 300 hours on AR213 and about fifty hours on other Spitfires. I first flew AR213 on June 12th 1975. Neil Williams broke me in on the aeroplane initially. The Yak-11 was the only other real high-performance aero-

plane I had flown and I had been flying that since 1973. Ever since I had been working on the Mk I, I used to taxi it around quite a lot. I had worked on it a lot and was familiar with all the systems. I had figured out how everything worked properly and how the emergency systems all worked, and I was familiar with running Merlins, so it was actually an easy transition. After the Yak, I found the Spitfire pretty easy

to fly. The first flight I did with it was fairly undramatic.

"My second flight in it was at an air show at Duxford, in formation with another Spitfire, a Mk IX. My third flight was a duo, with Neil Williams in MH434, then Adrian Swire's Mk IX Spitfire; we did the Pink Floyd concert at Knebworth Hall. It was July and it was one of those things where they were timing it. They were playing a

PRECEDING SPREAD: MH434 OF THE OLD FLYING MACHINE COMPANY, DUXFORD; BELOW: TONY BIANCHI AND THE MK1A SPITFIRE AR213 AT BERNAIS, FRANCE IN SUMMER 1984.

song called *Spitfire* and we were to break over the top with the two Spitfires as they started the song. Luton air traffic control relayed the time exactly when we had to run in and break; it was a split-second thing. That was the third time I had flown the thing and we did some aerobatics as well."

In more than thirty years of flying the Mk I, Tony has had his share of difficult situations. They mostly related to weather conditions, rather than to the aeroplane. "I used to go to France quite regularly with the Mk I to do air shows. In 1984, I had done a show in Paris for the D-Day celebrations and had a business appointment with a company at Bernais in

Normandy. I went in there on a Sunday night and left the aeroplane in their club hangar.

"It had been beautiful, crystal-clear June weather and was nice again the following morning. I didn't know there was a changing weather situation and, being June, you don't worry quite so much.

"Unusually, I couldn't pick up fuel at Bernais; they were waiting for their fuel delivery. I'd flown from just south of Paris and, in the good weather, it had only taken me twenty minutes, so I'd got forty minutes of 'reduced-power' fuel in the aeroplane and wasn't unhappy about setting off to go to clear customs at Deauville, which is a ten-minute flight from Bernais. I pre-

flighted the aeroplane, cranked it up and took off. I could see in the distance there were clouds. It was slightly darkening and I thought, 'well, I'll get to Deauville and I'll have a look at it there.' I didn't realise how fast it was shutting in. As I got five minutes out of Bernais it was starting to spit with rain and the cloud was building really fast; I'd never seen it build so quickly. I went down from 1,500 feet to about 800 feet and thought, 'God, the visibility is getting really bad. I don't like this very much. I'm just gonna turn around to the opposite way.'

"I would normally never take off in a Spitfire without having full fuel in it. One of the snags with all the early Spitfires, and some of the later ones as well, is you've got eighty-seven gallons of fuel, and probably seventy-five gallons useable, providing you don't chuck the thing around very much . . . you may even have eighty gallons useable, but ten gallons is nothing in the bottom of a big tank, so you don't ever risk it. You can't get a reading on the capacity of the fuel until it's on the bottom tank, and the bottom tank is thirty-seven gallons. The top tank is forty gallons and before I went off I unscrewed the thing and could see it hadn't got through the top tank, so it had got about forty+gallons in there, but you can't tell how much because there is no gauge on it; the gauge is on the bottom tank.

"So, I've got myself into a difficult situation without fuel. I'm so careful normally that I don't fly in bad weather. I've flown in some pretty atrocious weather in the Spitfire, but always knowing where I was and with plenty of fuel, and making for a point that I know is clear the other side.

"This was an unusual one. I didn't

bother to find out about it. It was my own fault. The bottom tank was starting to read that I had maybe twenty minutes safe. 'I'm gonna go back to Bernais and wait for the weather to clear or until their fuel is delivered. Maybe I'll stay another day; it doesn't matter.' I turned around and went back to Bernais and it was shut in completely there. The whole thing had just built up so quickly and there I was, sitting in pissing rain at 300 feet, pulling relatively tight turns and thinking 'I'm gonna get myself sorted out.'

"Fortunately, I had set the VOR for Deauville, but the trouble was, the VOR on the Spitfire is down on the floor on the right side and when you are flying down at low level and it's raining and you are unsure of your position . . . and it can get pretty hilly in that part of the world . . . you don't really want to look down to change frequencies or look to see where the needle is going. So, I was getting pretty worried by then.

"One of the things you do, which is probably a bit stupid, you put off the evil day by reducing the power. You slow the whole aeroplane up so you're not burning so much fuel. Then the engine doesn't run so nice, and that Mk I Spitfire always ran a bit rich if you started to reduce the power. Then I'd open the hood and, of course, get gassed by the carbon monoxide from the exhaust. I thought, 'I've got to make a decision.' I called Deauville and couldn't even get hold of them. 'It can't be that there's nobody there. Maybe it's because I'm flying so damned low nobody can hear me.' So, I went up into the clag a bit, to maybe 800 feet because it was absolutely thick, solid hard rain.

Men of Action need
BRYLCREEM
THE PERFECT HAIR DRESSING

Try the Active Service Packs

Larger bottles,
1/6, 1/9 & 2/6 **1'**

BRYLCREEM

TOP: THE SPITFIRE AB910 LANDING; RIGHT: THE SPADE GRIP AND COLUMN OF A CRASHED SPITFIRE.

"I was still heading in the correct direction for Deauville but didn't realise that I had probably got myself misplaced by orbiting. You orbit quite fast in a Spitfire; 140 knots or so while covering a lot of ground and using a lot of fuel. 'What the hell do I do?' I was actually lost. I had been orbiting, I was criss-crossing, and I got to the point where I thought, 'I'm really not with this. I'm really not sure what to do. Do I climb up and set a heading going out to sea, get up to 6,000 feet, put it upside down and bale out, because you can't land anywhere around there?' It's all typical Normandy small fields with stone walls. In England, you could probably find a field and stick it in.

"I kept calling Deauville and finally got through to them. I said to them, 'I'm an aeroplane bound for Bernais', and they said, 'No, it's impossible. We are giving thirty metres in heavy rain with visibility less than 500 metres.' I didn't know where the hell I was and then, luckily, I spotted a motorway. All the lights were on in all the cars and the road was going sort of east-west. I thought it must be the new motorway that routes around the south of Deauville. It was now absolutely pitch black, like flying at night. It was that bad, as it so often can be on the Normandy coast, especially where the estuaries are. I called Deauville again and said, 'I'm gonna try to get to you.' There was no response at all this time, almost like the place was closed down.

"Eventually, I saw a mast with a red light on it at about the level I was flying at, 300 feet or so. The visibility was slightly improved and I could see a ridge and just past it a hangar. I thought it must be Deauville and

decided to go for it. Then I could just see the runway, with which I was almost lined up. I put the gear and flaps down and was shaking like a leaf. I had been airborne for nearly thirty minutes by then and had been swanning around a lot trying to find out where I was. I didn't bother to call Deauville again; I just looked around to see that nothing else was coming in. I was coming up the hill and I landed on the end of the runway. I taxied on around and parked outside the control tower. I went in and said to the guy, 'The weather's bad!' He said, 'Did you park last night? Were you in the hangars over on the south side?' I said, 'No, I just flew in from Bernais' and he said, 'No, you haven't. They're giving twenty or thirty metres vertical here. You can't avoid the parking fee from last night. If you left it outside the hangar you still have to pay.' I said, 'I didn't. I've just flown in.' He didn't believe it.

"I had another situation in the early Eighties when I went into Headcorn in Kent for a Polish fighter pilots' reunion. That Dorking Gap is a lethal part of the world. Lots of people crashed there during the war, coming from south to north and getting caught in the gap. It's quite high in there, about 800 to 900 feet. You've got the flat of the south of England; then you follow the railway line up and there are all those hills. It wasn't that bad this day, but then it started to rain quite hard. I thought, 'Well, I know the area well. I don't really need a map in those parts.' Usually you just fly up past Red Hill. You don't go through the hills or you start to get involved with the London control zone. You route using Gatwick and come up past Red Hill and head

for a mast at Bagshot which is between the control zone of Farnborough and the southwest corner of the London control zone. I couldn't even see the mast but thought, 'It'll be all right. I'll press on.' Then, fortunately, I saw the base of the mast. I was down around 200 feet and it was getting pretty grim. I decided I wouldn't work London control zone because they would know that I was flying too low.

"I shot past Woking and lost it completely. I couldn't see a thing. It was just sitting in the trees, all the mist and rubbish. I did a great circle and thought I would go and land at Blackbush. It was the worst I had ever been in, to the point where, if I had pressed on I was going to crash. I did a couple of 360s and confirmed that it was sitting in the trees everywhere. 'Right. I'm gonna go on my original heading because I've got nothing else to do. I'm gonna find a field and crash. If I can find Ascot Race Course, I can land there.' I started to do a big turn. Ascot is in the London control zone, but I was so low I figured they wouldn't pick me up anyway. 'God, nothing is around here.' I did this huge turn, couldn't find Ascot and headed off on a northerly heading. I was dodging around the bigger trees and kept thinking, 'What am I going to do?'

"Then I picked up a village that I knew was between Bracknell and White Waltham airfield. I thought I could follow the road up but worried about running into one hill that I knew was on the north side of White Waltham . . . everything else was pretty flat. I was down to about 200 feet, in and out of the trees, when I found White Waltham. I don't know how I did it. I went right round the airfield

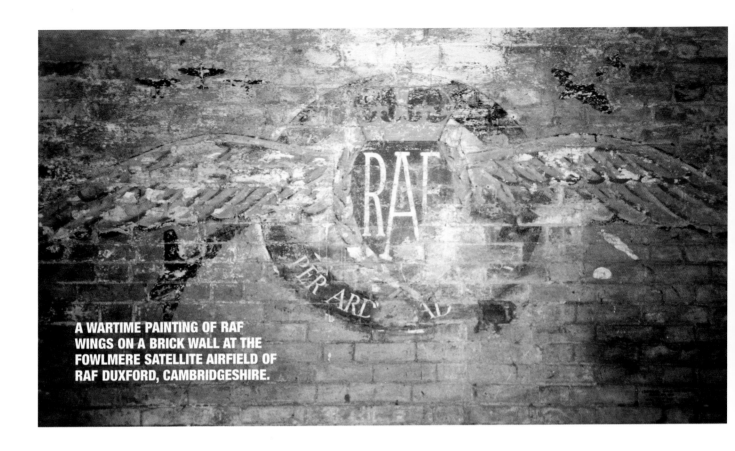

with a wing down. Somebody said it looked like a one-winged Spitfire going around the airfield. One wing was sitting in it. I was doing a gentle turn and one wing was twenty feet from the ground. I was younger then and used to flying things like Stomps and the slower biplanes. You get used to flying around in crappy conditions. You press on and the weather changes very slowly. You press on in fighters like the Spitfire and the weather changes very quickly and you get committed. With aerobatic aeroplanes you can at least dump them in a field. You may risk turning one over on its back. The limitation was that, if it was so bad and you took off in it, then it's your own fault, but if it actually gets bad en route, then you turn back, and I was always a great one for turning back. But if there's nowhere to turn

back to, you're stuffed."

"I used to do quite a few shows, both in AR213 and in a Mk IX, in the Normandy and Paris areas. In those days there were unlimited Spitfire shows to be had in Europe. Nobody else was really doing it in the '80s and I used to do a lot of shows, going down there as both engineer and pilot. I would find a few people I had met before on other show sites, French engineers, who would say 'If you are going to this place near Paris I will come and meet you and I will bring my tools and help you.' They were some really good engineers who would help me do the pre-flights or if I had any snags. With help from them I would do things for myself and leave the aeroplane over there. Two weeks later I would go back for

another air show and then bring the Spitfire back to the UK. It was really good fun operating totally on my own. I never even wore a flying suit in those days, just a helmet, a shirt and jeans. When it was really hot in summer I used to fly around in shorts and a tee shirt."

RAF Wing Commander Alan Geoffrey Page finished the war credited with seventeen aerial victories. He was badly burnt when his fighter was shot down off the English coast during the Battle of Britain, endured many subsequent operations to repair the damage to his face and hands, and successfully returned to combat flying for the duration of the conflict. He remembered his first flight in a Spitfire. "I had been sitting in the cockpit for half an hour memorising the procedures for take-off, flight and landing. An airman climbed onto the wing behind me to help me with my parachute and harness. Word spread swiftly that an unusual first solo on type was taking place and soon ground and air crews were gathering to watch with morbid interest.

"As soon as I was properly strapped in, the squadron commander climbed onto the wing for a final word. 'Don't forget, taxi out quickly and turn her into the wind—do a quick check and then get off. If you don't, the glycol will boil and so will my blood. Good luck!'

"I responded with a nervous smile, closed the tiny door and turned to face the mass of dials, buttons and levers. For a moment panic seized me and the temptation to undo the straps and get out was very great.

"The enquiring voice of the airman standing by the starter battery reminded me of the engine starting procedure, and my nervous feeling passed with the need for concentrated action. Throttle about half an inch open—gas on—nine full strokes in the KI-gas hand priming pump for a cold engine—propeller in fine pitch—brakes on—stick held back—press the starter button. I raised my thumb, the waiting airman replied with a similar sign, and I pressed the starter button firmly—the propeller began to rotate . . .

"A trickle of sweat ran down my forehead. Suddenly the powerful engine coughed loudly, blew a short stream of purply-white smoke into a small cloud and roared into life. Remembering that I had little time to spare before the temperature reached the danger mark of 110°, I waved my hands across my face. The waiting airman quickly ducked under the wing

and pulled away the restraining chocks. Glancing down, I was alarmed to see that the glycol coolant temperature had risen from zero to 70°. Releasing the brake, I eased the throttle open and the surge of power carried the aircraft forward rapidly over the grass.

"Was everything ready for a quick take-off? I wondered. I figured I'd better call up Flying Control and get permission to take off immediately. Pushing over the switch on the VHF box, I tried to transmit. 'Idiot!' I said to myself, switch the damn thing on. Another glance at the temperature showed 95° and still a long way to go before turning into the wind. The radio came to life with a whine, and contact was made with a fellow human being. The controller's voice was soothing and for the first time since strapping into the narrow cockpit, I relaxed slightly. But I was still none too happy.

"The temperature now read 105° and there were still a few yards to go, plus the final check. Softly I prayed for help. Temperature 107° . . . Now for the drill: R.A.F.T.P.R.—the radiator—God alone knows how many times I'd vainly tried to open it beyond its normal point to try to keep the temperature down. A—airscrew in the fine pitch—that's okay. F—flaps. Temperature 109°.

"I abandoned the remainder of the cockpit drill and, opening the throttle firmly, started the take-off run. The initial kick from the rapid acceleration drove the worry of the engine temperature away for a while. Working the rudder hard with both feet to keep the sensitive little machine straight, I was too busy for other thoughts. Easing the stick for-

ward, I was startled by the rapidity with which she responded to the elevator controls. The long nose in front of me obscured the rapidly approaching end of the airport, but by looking out at an angle, I was able to get an idea of how far away it was. If the glycol boiled now at this critical stage, the aircraft would be enveloped in a cloud of white smoke that would prevent me from seeing the ground when the inevitable engine seizure and crash landing followed. Looking back into the cockpit again, I saw the hated instrument leering at me. 110°.

"Accompanying the feeling of fear was a new sound. The wheels had stopped drumming and a whistling note filled the air. The Spitfire soared gracefully into the air, thankful, as I was, to be away from her earthlybonds.

"Inside the cockpit I worked desperately to get the undercarriage raised. The C.O. had explained to me that because the starboard aircraft leg hung down in front of the radiator when the wheels were lowered, this affects the cooling effect by the airstream. By raising the wheels the air would pass unhindered through the radiator to do its work. But here I fell into trouble again. To raise the wheels, I had to move the selector lever and this was on the right hand of the seat. I then had to pump them up with twenty movements by a long handle, also on the right. To do this while flying the very sensitive aircraft meant using my left hand for the control column, while the right hand struggled with the undercarriage mechanism.

"The Spitfire was now about twenty feet up, gaining speed rapidly and skimming over the trees and hedges. I selected 'wheels up' and gave the han-

JONATHON WHALEY TAKES AR213
FOR A POST-RESTORATION FLIGHT.

dle a first stroke. The engine cut out for an instant and the nose plunged earthwards. Being unused to the technique of keeping my left hand absolutely still while the right one moved forward, I had inadvertently pushed the control column forward simultaneously with the first pumping stroke, thus causing the machine to dip suddenly. The negative G placed on the carburettor had caused a temporary fuel stoppage. Some trees flashed by alongside the aircraft as a frightened pilot hauled back on the stick, and soon I was soaring skyward again, pumping frantically after removing my left hand from the control column. At this stage it was obvious that the Spitfire could handle herself better than I could. After this nightmare, the green light finally shone on the instrument panel, indi-

cating that the wheels were in the locked up position, and the engine temperature gauge showed a healthy fall. I took a moment to utter another silent prayer, this time of thanks.

"Now I had some breathing space, so I was able to look about and concentrate on the other aspects of flying the aeroplane. Throttling back the engine and placing the propeller in coarse pitch, I allowed myself the luxury of relaxing slightly and looked down on the beauties of the Norfolk Broads. However, the pleasures of the English countryside didn't last long. Glancing down and behind me, I was horrified to discover that the airport was nowhere in sight. The swiftness of the Spitfire had soon taken me out of sight of the landing ground, and although homing facilities were available over the R/T, pride stopped me

from calling the flying control tower for assistance. Instead, a worried young man flew about the sky in circles anxiously peering down for a sign of home. Ten minutes later, relief flooded through me when the unmistakable outlines of Norwich Cathedral appeared out of the summer haze, and from there the airport was easy to find. A minute later the graceful plane was banking round the circuit preparatory to landing.

"Again I recalled the cockpit procedure given to me by the squadron commander: R-U-P-F—radiator, undercarriage, pitch and flaps. This time the pumping down of the wheels came quite simply, and the other essential procedures prior to the final touchdown followed. The exhaust crackled delightfully as the engine was throttled back and the

BOB DOE SERVED WITH NOS. 234 AND 238 SQUADRONS, RAF; RIGHT: THE OLD MAIN RUNWAY AT RAF TANGMERE, SUSSEX.

plane came in gliding fast over the boundary hedge. In the cockpit, I eased the stick back and the long streamlined nose rose up and cut out the forward view of the landing run. Looking out to the left, I carefully judged the height as the Spitfire floated gracefully a foot or two above the green grass, then losing speed she settled down on the ground to the steady rumble of the wheels. As soon as the machine had come to a halt I raised the flaps and thankfully undid the tight-fitting oxygen mask. The pool of sweat that had collected trickled down my neck. With a newly born confidence, I taxied the machine back towards the waving airman near the hangar. Just as I removed my helmet and undid the confining harness and parachute straps, Dizzy Allen walked up. 'Back in one piece, I see. How'd

you get on?' Trying to appear nonchalant, I replied, 'Loved every minute of it. She certainly handles beautifully.' The feeling of achievement obliterated the memory of the fear I'd felt during most of the flight, and now I felt justified in taking a place among my fellow fighter pilots."

Another man who has frequently flown AR213 is Jonathon Whaley, who also owns and flies a Hunter F58a jet fighter. He describes aspects of flying the Mk Ia, while she was still fitted with the Merlin 35 engine and Rotol four-bladed variable pitch propeller. Jonathon wrote the following shortly before AR213 entered the lengthy restoration, down to the last rivet and back up again. "Before the work started, I made several flights to record her handling and performance

in order to be in a position to make an objective comparison with her post-restoration state. The dedication and obsessive attention to detail of the team carrying out the restoration has been something to behold. Thus, the responsibility on those who will fly her in the future is immense. Spring 2007 and her restoration is nearing completion, so I look forward to the days when I will fly her again and make the comparisons, all in the cause of enlightenment of course and any grin on my face, coincidental.

"Anything labelled a Mk Ia could be taken as meaning Mark One, Mod. Zero, implying something one up from the prototype and better things are expected. From a pilot's point of view, with the proviso 'with this particular Spitfire Mk Ia', I don't believe this to be true. I've been privileged to

have flown Spitfires Mk Ia, Mk IX and Mk XVI. The latter two, with their owners' very generous permission to fly them for my own pleasure rather than limited to airshow displays. I have to be up front and say that AR213 has a Merlin 35 with a Rotol four-bladed variable pitch prop rather than an original fixed-pitch two-bladed prop or later two-pitch position prop, nor have I flown a Mk II or Mk V.

"From Mitchell's design, his team got the production design right from the start. From a pilot's perspective, all further developments took the Mk I further from simplicity and perfection in the drive to increase power, carry bigger guns, go higher and fly longer. Without question, the later Mks were superior fighting machines, but that particular hell is past and I'm not at war.

"So what is she like to fly? In the world of piston aircraft, it is all your aviation dreams, desires and fantasies rolled into one, with enough little 'gotchas' to add piquance to the actual, unadulterated, shameless fun of flying her. It simply is not possible to convey in words the kick one gets and addiction from being in the air in an aircraft that in an Eartha Kitt purring voice says 'I'm yours, you're mine, you'll never want to leave me.' I know *Miss Demeanour* (my Hunter F58a) gets jealous, but then again, my wife gets jealous of both of them! If I could only fly one Spitfire, it would be this Mk Ia.

"Although AR213 has very low hours, she has never undergone a full restoration. Moreover she had, in her early days, a couple of mishaps requiring a 'return to manufacturer' for repairs. Her current owner is, I am pleased to report, slowly working through a rolling rejuvenation pro-gramme under Tony Bianchi and Personal Plane Services Ltd at Booker. So far, the flaps, ailerons and tail section have received attention. In the pipeline are the fitting of her three-bladed DH prop and Siamese exhausts. (Although acoustically I prefer the single exhausts which do not have such a tinny sound, and at the risk of offence, like a Hurricane.) There are also plans to fit the original-style flat-sided hood.

"The walk-around is much like any aircraft, although perhaps a more detailed inspection is made on individual panel fasteners both on cowlings and wing panels. Parking on 'hard' and taking off from grass are the preferred options. Any leaks of coolant can be spotted on the hard, but when parked on grass, are virtually undetectable. Many of the other items given special attention are not mentioned in any

Pilots Notes but are learnt from talking to engineers and other pilots based on their own experiences. One such item is the undercarriage down-and-up lock pin. Is it well greased and correctly orientated? If not, the undercarriage may be difficult or impossible to retract or lower.

"The pilot's seating position is rather like sitting on a kitchen chair rather than a modern fighter's semi-reclined seat. There is a gesture to increasing G tolerance by having a 'top' step on the rudder pedals. A seat-pack parachute is worn and a simple four-point harness straps you in. Pre-start checks consist of full and free movements and then, working left to right, rudder trim for movement and neutral-setting or last-flown position, elevator trim movement and set one-down, throttle movement and cracked open, pitch movement and

fine, friction set (piquance later), radiator operation and set open, battery master on, Comms a/r, mags off, flaps up, instruments checked, fuel on, prime (piquance later) and lock, u/c lever visual check, emergency u/c down lever wire lock unbroken, and finish with opening the compressed air feed for brakes and check pressure. AR213 does not have a self pre-oil capability which, therefore, requires the engineers to have hot pre-oiled the engine within the previous five days. This puts oil in the galleries and reduces the time on start-up for oil to reach the far flung points within the engine, keeping wear to a minimum.

"A look around outside and apply the parking brake which consists of locking the bicycle-type brake lever mounted on the control column. If you've remembered the air valve, a satisfying 'passing wind' sound

escapes. If not, you've either forgotten the air valve or there is no air in the system, which will mean no brakes until the engine driven pump has built up pressure (piquance later).

"A meaningful shout of 'Clear Prop', hold the stick back with a combination of knees and elbows, left hand hovering around the mag switches, right hand index finger over the booster coil button and middle finger pushing on the starter button. Three or four blades later, the booster coil button is pressed, which, if you've primed correctly, will result in a satisfying blast that only a BIG V12 can produce. If not, either nothing happens or the view outside gets interesting. Gentle but copious orange and yellow flames start emanating from the exhausts and lick their way down the fuselage sides. I've seen (but not experienced) them reach the cockpit.

With flames, now is the time to keep metaphorically cool. Whatever happens, keep the starter button pressed and the booster coil active. Failure to do so will result in a carburettor fire of almost certain catastrophic proportions. Priming from cold (four good pumps) has never given me anything other than a clean start. It is priming a hot engine that requires 'a feel'. The priming pump itself needs to prime before it pumps properly. The expression 'one and a half good pumps' is often used but what amounts to less than or better than 'good' is not always clear until after the event. I vividly remember my first wet start—as normal, goofers in abundance—those few in the know grinning at my predicament and those not, pointing wide-eyed and shouting 'Fire'!

"Once the engine starts, the mags are flicked on and the booster coil released. With the Merlin 35, if the mags are on at the start, the engine will still fire up, but not quite as cleanly. On the bigger Merlins, trying to start with the mags on will result in harsh kicking back as ignition takes place too early at the turn-over speeds.

"The oil pressure rises rapidly and

ABOVE: THE WHITE HART AT BRASTED, FAMOUS HAUNT OF THE AIRMEN OF RAF BIGGIN HILL IN WWII; BELOW: A SPITFIRE IN THE IMPERIAL WAR MUSEUM.

you ease up to 1000-1100 rpm. If this is your first Merlin start-up, from within the cockpit the power-train at 800 rpm sounds like it is falling apart. If your car engine and gearbox sounded like that, you'd shut down before it all stopped with a bang. All this clatter smoothes out when the engine comes under load.

"If the start has been made with a warm engine and there was no air pressure before the start, either due to a leaky system or because you forgot to shut off the air valve on shut down, there is now a race on to see what is reached first—enough brake pressure to taxi and take off, or coolant boiling point. From a cold start, on a warm summer day, you have a leisurely five minutes, give or take a second or two, between starting and must-be-airborne. If the engine is already warm, it is a case of starting, pre-take-off checks on the taxi, line up, mag check and prop exercise all within three minutes or less. At busy airfields, pre-briefing ATC and requesting permission to Start is an essential pre-flight action. The phrase 'a watched kettle never boils' ain't true. This one boils between blinks. (For the experts, the radiator is a clean and clear one.) Unlike the later marks, the Mk Ia has an unrivetted header tank and therefore is not as pressurised, resulting in venting coolant as 100°c is reached. The best way to safely pre-experience the taxiing and landing views is to fly a Harvard from the back seat. Apart from the lack of visibility, taxiing is easy, weaving as required to see in front. The attitude to take is akin to wheels-up landings; there are only pilots who have taxied into something and those who are going to.

"Pre-take-off checks are traditional, although going above 1700 rpm for mag checks and prop exercising takes you in to the realm of tipping her on her nose. On the Mk XVI, I recall feeling, only just in time, the tail become light. For take-off, from a safety aspect, it is wise to leave the hood locked fully-open. In the event of tipping over on your back, for whatever reason, you've an exit. The danger lies in the hood jamming closed on the rails.

"If, for some reason after the start, you've been delayed and the coolant temperature has continued its inexorable climb, you have to do some serious risk assessment. If you can get airborne, even as it reaches 120°c, before you have the gear up, it will have cooled to 100°c or less. If, on the other hand, you gamble and lose, when you shut the engine down it will heat-sink the temperature to an even higher figure. 105°c with steam boiling out her side, is my decision point.

"By now you might be thinking 'hold on, he said she was all your ideals, desires and fantasies rolled in to one, with enough little "gotchas" to add piquance to the actual unadulterated fun. Flames and boiling coolant don't sound like fun!'

"AR213 is, at 5058 lbs basic and with 1240 bhp on take-off, about as good a power to weight ratio as you can make a Spitfire (Typical Mk IX 6210 lbs and 1665 bhp). What is more, as opposed to the later marks with bigger engines, once you've experienced the power, you can release the brakes and pile on the full +8 boost with a smooth determination and ear to ear grin. Nirvana here I come! With the hood open, the Merlin roars in the ears no matter what head gear you wear. The acrid

exhaust pours momentarily into the cockpit and the tail comes up quickly with the stick moving forward, somewhere between leading and following. Your first brief view ahead arrives just as the stick can be teased back to shake the ground clear. On a bumpy grass runway such as Booker's 17/35, from tail off the ground, it's only a couple of bounds and you're airborne, almost as if the aircraft is kicking the ground clear. Brakes on/off and now you change hands, putting your left hand on the stick and right hand on the u/c lever. Throttle friction tight enough? No? Then swap back hands, put the power on again, tighten the friction and start over! Raising and lowering the undercarriage is perhaps the only murky area in flying a Spitfire. I won't bore you with the technicalities, but suffice it to say that I doubt if there are many Spitfire pilots who have not 'beaten the gate'. This results in the lever being in the raised position with the gear still down or vice versa. The secret is to be purposeful and without haste in all movements. Pull, one Hail Mary, rotate. If the lever sticks, it may be necessary to invert the aircraft (where the engine will stop) or give a prolonged push to unload the pins and allow the lever to be rotated. If the gear stays down, the coolant will almost certainly boil at cruise power.

"With the gear on its way up, take some of the power off and reduce the pitch initially to 2700 rpm, looking for 120 kts to start climbing. Climb speeds are academic since the heights normally used are reached in seconds by which time, if you are transiting, you can pull back to between 0 and +1 boost and 1800 rpm. This will settle you down to just over 180 kts.

A SPITFIRE SERGEANT PILOT READY FOR TAKE-OFF IN THE BATTLE OF BRITAIN, SUMMER 1940.

All writers say the controls are superbly loaded and harmonised and if you want to be super-critical, the elevators are on the twitchy side. Why should I be any different? On a cool day, the radiator flap may now be moved into the trail position for a transit. I've only once needed the closed position, during a cold winter day. Then I wished she had the same heating facility as the Hawker Fury, with its RR Kestrel and belly radiator from which air could be deflected up into the cockpit.

"Slowing down, the stall is benign with good control authority all-round, and provided you don't pile on the power in the recovery, very straight forward. There is a good buffet warning and a wing will drop gently. Pile on the power here to get an idea of how terminal a mistake it would be to do so on finals. Spinning is not permitted (was in the Service) but I have the feeling that even with the Merlin's weight and 4-blade prop adding to the gyro forces, recovery would present no problems.

"For aeros and displays, setting 2500 rpm and using up to +4 boost gives you all the power needed. AR213's four blades and CSU are not well matched. The CSU is unable to fine off enough as the speed increases and at 300 kts the limiting 3000 rpm is reached. I have an inventory of display manoeuvres for the Mk Ia and the maximum I need is 260 kts for a pull-up to vertical 1/4 or 3/4 roll (latter always left) and pull through. The start of this manoeuvre also sees the maximum 4 g, otherwise 3 g suffices. For displays I never do a pure loop and the roll in the vertical ensures a gently played-out pull-through follows. One of the most

pleasing manoeuvres to fly is a half Cuban, making the half-roll on the way down, smooth and prolonged. Any temptation to make it a Competition-type straight 45° should be avoided. The Merlin does not like anything other than positive G and slowing the roll removes the obviousness of the barrel. Provided power is not being applied at the time, and not from speeds under 150 kts, rolling right has no more considerations than rolling left. If, on the other hand, you're only 25° nose up, powering up and you've only got 120 kts, you'd better make sure your Derry turn is to the left. A stall turn would be easy to carry out to the left but I've never tried and don't believe it is appropriate to the type. However, a steep, past the vertical, ballistic wing-over with a tweak of rudder, can be very graceful, giving the crowd a perfect plan-view of the distinctive elliptical wings. A Victory Roll is mandatory for type.

"During a display, the coolant temperature will steadily rise such that on a hot day, 90°c will be seen. If this is the case and you intend to land straight off the display, you've a problem ahead. It is wise to plan the end of a display with a relatively high-speed low-level pass with just positive boost, allowing you to break downwind, throttling back, all while gently cooling down. The hood is opened again for safety and, if I have time, half-unlock the cockpit door. I aim to fly a constant turn on finals, rolling wings level just before flaring. The u/c and then flaps are lowered just as the finals turn and descent is started; here I like to see 95-100 kts. Delaying u/c and flaps helps the cooling-down process by not disturbing the airflow into and

out of the radiator. This makes the early finals turn quite busy, but the essentials are simple; a green DOWN light (AR213 has no wing mounted u/c indicators), flaps down, brakes on/off exhausted, air pressure up and pitch fully fine, Ts & Ps.

"Power On or Off approaches can be made, and to be honest, I take each approach as it comes. If I have power on, then I can come back to 75 kts, 80 kts if power is off. If you find yourself in a power-off approach, only continue if the approach is looking good and there is no prospect of a sudden go-around(traffic ahead?). You should have cooled down to the low 80's by touchdown, because now the temperature will start climbing again. Ideally, the flare will see the speed falling back through 60kts just as you touch. On a bumpy grass runway, with 60+ kts, you'll bounce and AR213 loves to stay airborne. Just be patient and she'll gently but reluctantly give up flying. On occasions where I have been on the fast side, I have raised the flaps just as I've touched. The flaps are so slow to come up, that whether or not raising them prevents the bounce and float, I don't know. Maybe someone will tell me that the drag from leaving the flaps down reduces the landing roll, but I'd want coloured graphs to prove the level of significance, and if you're fast and that tight for space, a go-around should have been made. The only other consideration would be that compressed air is being used and if there was a problem with the flaps air system, you'll deplete brake-pressure air.

"The rollout in a cross-wind can be interesting and any temptation to use the brakes on a hard runway must be ignored. The secret is to keep flying the aeroplane and not treat it as fast taxiing. Once comfortable, flaps up, check radiator open and look for the quickest way back to the dispersal. Weaving back, it is perhaps time to reflect again that you're in a very expensive, extremely rare, flying Mk Ia and have once more taken a large chunk out of your allotted share of life's good fortune. Back at the pan, the power is reduced to idle and an idle cut-off ring is pulled to stop the engine. With mags, fuel, comms and battery off, you've only to remember to turn off the air to conserve the pressure. In the background there will probably be a 'wet fart' sound as steaming ethyl glycol vents forth in a satisfied manner from the pipe in the starboard nose cowling, the kettle's boiling."

Wing Commander John A. Ward wrote of his most memorable Spitfire moment: "Born at the beginning of the jet age, my passionate love of aeroplanes and flying began when I progressed from Blyton to Biggles and it didn't take very long to realise that I had been born thirty years too late. I had to come to terms with the fact that, by the time I was old enough, the Spitfire would be only a museum piece; just a legend, long gone, and flying one was just a wild boyhood fantasy. I went on to fly many other fighters, until I found myself in the undreamed-of position of being offered a job where I also had an opportunity to participate in Spitfire display flying.

"I have been asked to describe my most memorable moment in the Spitfire. My difficulty is in finding the words to convey the feelings and emotions of that moment adequately. I

hope it will help if I set the scene; a few words about the events that led up to that moment may help.

"It was a Sunday, 15th September, the day that the Nation remembers the Battle of Britain. The plan was for a Spitfire flypast over the British Military Cemetery at Hermanville, a small village in France, roughly midway between the beaches of Normandy and the town of Rouen. Sir Harry Secombe was leading the BBC *Songs of Praise* commemoration at the little cemetery, which was surrounded by a low wall and shaded by a clump of trees. I made the flypast, dipped wings in salute and then came back around for the upward victory roll over the assembled gathering.

"As I headed back across the French countryside towards the Channel and to Bournemouth to refuel, I reflected on the fact that, before I was born, this very same Spitfire that I was flying had fought in the Battle of Britain, and had been part of the same war in which the men in that cemetery had fought and died, and I wondered how many times in that war she had made that return flight back home across the Channel. It was a sobering moment that made me feel very humble. I was simply the means whereby this thoroughbred old warhorse had saluted fallen comrades in a simple gesture of thanks and remembrance. I flew northwards across the Channel as so many Spitfire pilots had done in the war. I could see The Needles just right of the nose, bathed in bright sunlight—a welcoming sight which seemed to call, 'This is England; home is this way.'

"After refueling at Bournemouth I took off again, heading northeast for home. It was late afternoon, but the

sun was still strong and bright. Traces of high cirrus cloud gave an impression of space to an otherwise deep blue sky. Settled down at around a thousand feet, with the South Downs off the right wing, everything suddenly seemed surreal. Apart from the sweet comforting song of the Merlin, there were no other sounds or movements at all; it seemed as if I was alone in the world. The radio was silent, and the sky was empty. Ahead and on each side, as far as I could see, lay the English countryside. My world seemed to be standing still. In a perfect and memorable moment, I felt an integral part of that superb machine. I was totally immersed in the thrill of being swept across the woods and fields of England, the Spitfire eager and instantly responsive, as though she was straining to rush me headlong towards the distant horizon. My boyhood dream had come true. It was my 'Spitfire moment' and I wanted it to last forever. The memory will be with me always."

STATIONED WITH 12 GROUP, RAF IN 1940, SERGEANT GEORGE UNWIN WAS CREDITED WITH FOURTEEN ENEMY AIRCRAFT SHOT DOWN BY THE END OF THE YEAR; OVERLEAF: A T MK 9 TRAINING SPITFIRE POSTWAR.

A CANNON-ARMED SPITFIRE READY FOR TAKE-OFF ON A STRAFING SORTIE.

SPITFIRE IN WW2

"At that moment I saw dimly a machine moving in the cloud on my left and flying parallel to me. I stalked him through the cloud, and when he emerged into a patch of clear sky I saw that it was a Ju 87.

"I was in an ideal position to attack and opened fire and put the remainder of my ammunition—about 2,000 rounds—into him at very close range. Even in the heat of the moment I well remember my amazement at the shattering effect of my fire. Pieces flew off his fuselage and cockpit covering, a stream of smoke appeared from the engine, and a moment later a great sheet of flame licked out from the engine cowling and he dived down vertically. The flames enveloped the whole machine and he went straight down, apparently quite slowly, for about five thousand feet, till he was just a shapeless burning mass of wreckage.

"Absolutely fascinated by the sight, I followed him down, and saw him hit the sea with a great burst of white foam. He disappeared immediately, and apart from a green patch in the water there was no sign that anything had happened. The crew made no attempt to get out, and they were obviously killed by my first burst of fire.

"I had often wondered what would be my feelings when killing somebody like this, and especially when seeing them go down in flames. I was rather surprised to reflect afterwards that my only feeling had been one of considerable elation—and a sort of bewildered surprise because it had all been so easy."
—Flight Lieutenant D.M. Crook, DFC, 609 Squadron

In the very first combat engagement

of the war, RAF Fighter Command's 602 and 603 Squadron Spitfires from Drem and Turnhouse respectively, and Hurricanes of 607 Squadron from Usworth bounced a flight of Junkers Ju 88 bombers of 1/KG 30 out of Westerlund on Sylt. The German bombers were attacking Royal Navy targets in the Firth of Forth, and their crews were under the illusion that the targets were only defended by a small force of Gladiator biplane fighters. In the action, two of the Ju 88s were shot down.

Contrary to the belief of thousands of British Expeditionary Force soldiers who were evacuated from the beaches of Dunkirk in May 1940, the Spitfires and Hurricanes of Fighter Command were present and doing their job in the skies of northern France. On May 25th, Spitfires attacked fifteen Ju 87 Stuka dive-bombers on their way to strike at the docks and the troops on the beach. The Spitfires had found the Stukas over Calais and destroyed four of them. During the remainder of the massive evacuation exercise, the Dunkirk airspace was continuously patrolled by no less than sixteen Spitfire and Hurricane fighter squadrons of Eleven Group, Fighter Command. On the 26th, the British pilots downed twenty of the enemy aircraft. The actual score by the end of the nine-day Dunkirk evacuation was essentially equal, with the RAFCV and the Luftwaffe each destroying about 170 enemy aircraft. But the victory must be credited to the British who showed the Luftwaffe its vulnerability in the air.

After Dunkirk, the Luftwaffe struck Britain in a number of night raids, including one on the Southend area of Essex. On the 18th of June, the wife of 74 Squadron pilot Adolph "Sailor" Malan had returned to their Southend home after giving birth to a son, Jonothan, a few days earlier. For several evenings Sailor had been kept awake by the incessant pounding of the local anti-aircraft guns firing at German raiders. One evening he decided to ask permission to take off on a night patrol in his Spitfire. With no time to dress properly, Malan's rigger and fitter quickly donned their gum boots, grabbed rifles and their tin hats and presented themselves for duty at the Spitfire dispersal, still in their pyjamas.

They rushed to start the Spitfire while Malan fastened his parachute harness. He strapped himself in and opened the throttle to warm up the engine. Overhead, he spotted a Heinkel 111 bomber at about 6,000 feet, coned in searchlight beams. It was headed straight across the airfield in his direction and, realising his predicament he unstrapped and leapt from the plane, making a dive for a nearby trench. He thought that the trench was about eighteen inches deep but the base personnel had continued to dig it out recently until the depth had reached five feet. As the Heinkel crossed overhead he plunged into the ditch and landed face first in the mud at the bottom. As soon as the bomber passed over, Sailor emerged and climbed into the idling Spitfire. He made chase and soon intercepted the bomber as it headed for the coast in a slow climb, the crew blinded by the searchlights.

Sailor later described the Heinkel as looking like a moth in a candle flame, until its wings took shape and he suddenly realised how close he was. He signalled the [anti-aircraft] guns to stop firing and attacked, pressing the gun button. After a three-second burst he had to jam his stick forward to avoid collision with the enemy. The windscreen of the Spitfire was immediately covered with oil from the Heinkel. The enemy bomber arced away from the searchlights and spiralled onto the beach, half in and half out of the water.

Returning to the Hornchurch base, Malan sighted another He 111, also coned by searchlight beams. Climbing towards the German plane, he again signalled the anti-aircraft gunners to hold their fire. At 16,000 feet he reached the Heinkel and approached this one with more caution, determined not to risk overrunning it at a range of 200 yards. He began firing from 100 yards. As he closed on it, the bomber became enveloped in flame and entered a steep spiral descent. It crashed in the garden of a vicar's home near Chelmsford, Essex. The resulting blaze was visible in Southend. When he arrived back at Hornchurch, Sailor hurried to a telephone to call his wife. She and their baby had slept through the entire enemy raid.

With the start of the Battle of Britain, the pilots of Mk I and Mk II Spitfires began to learn the limits of their weapon's capability. Supermarine chief test pilot Jeffrey Quill had his hands full testing new production aeroplanes at the Eastleigh factory near Southampton, but felt strongly that he should somehow get into the front-line action. He knew the company would object but he believed that in order to do his job properly, he had to have some first-hand experience of air fighting in a Spitfire. He wangled an assignment with 65 Squadron at RAF Hornchurch, Essex, in August 1940.

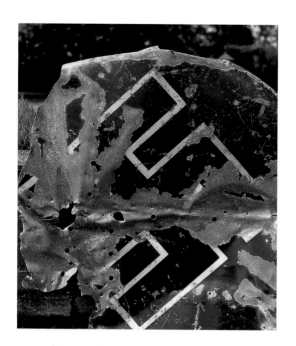

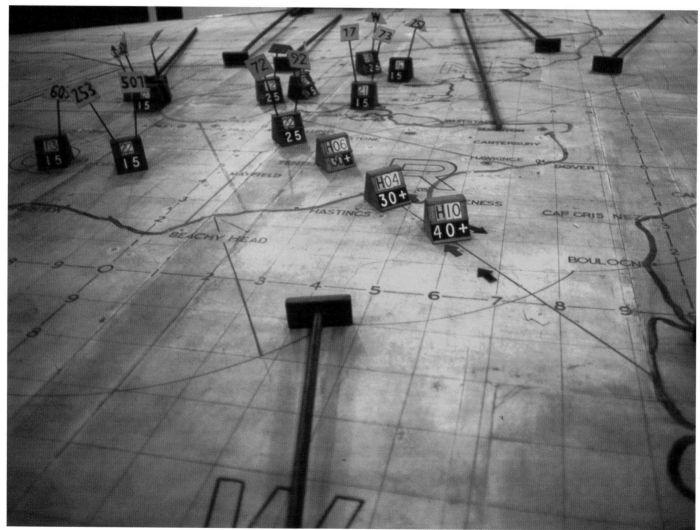

At that time the station was home to a number of Battle of Britain luminaries including Al Deere (54 Squadron), Sailor Malan (74 Squadron), and Paddy Finucane (65 Squadron). Quill was exposed to the fundamental tenets of air fighting: "Get in as close as you can before firing; you're usually further away than you think", "If you hit a 109 don't follow him down to see him crash—another will get you while you are doing it", "You need eyes in the back of your head", "Scan the sky constantly—it's essential you see them before they see you", "Never get separated if you can help it—and don't hang about on your own."

Of his first combat patrol, Quill recalled: "When our section closed up behind the rest of the squadron and became an integral part of twelve Spitfires, alive and animated and glistening in the summer sunlight, I suddenly felt enormously elated and excited. I had been flying this aeroplane for over four years; four years of wondering how it was really going to work and acquit itself when the real test came, four years of reports and meetings and discussions and continuous flying, day in and day out. But, looking back, it all seemed to have been one stage removed from reality, as if I had in a sense been a spectator or a planner rather than a participant; a coach rather than a player. Now, as we headed south-east across the estuary for Manston at a brisk clip and in quite tight formation, I felt a participant at last, and experienced a great sense of fulfillment.

"Nearly all our many engagements with Me 109s took place at around 20,000-25,000 feet. We had the edge over them in speed and climb and particularly in turning circle, but there were one or two glaring defects with the Spitfire which needed urgent action. I used to talk with Joe Smith on the telephone or write short notes with promises of a full report later. One point was the immense tactical advantage accruing to the side which saw an enemy formation first and in good time, and conversely the rapidity with which one was in trouble if one failed to see them first. The Spitfire was fitted with a thick, armoured-glass panel in the centre of its windscreen but the side panels were of curved perspex and the optical distortion from these made long-distance visual scanning extremely difficult. I was determined to have the design altered and indeed I succeeded in getting optically true glass into the side panels by 1941."

On August 24th, 65 Squadron received a signal from Eleven Group instructing Quill to return to work at Supermarine where his services were urgently required in the testing programme of the new Mk III Spitfire. He was just getting adjusted to the routine and procedures of the squadron, but with plenty more to learn, and was irritated by the order. He used his remaining hours on the squadron to consider what recommendations he could make for improvements to the Spitfire, based on his brief but intense combat experience. Most urgent, he thought, was a substantial improvement to the aileron control at high speed. Next in priority was the optical quality of the windscreen side panels, and the need for rearward vision improvement. There was also the matter of the engine cutting out under negative G.

The aileron problem was severe.

When the first production Spitfires were built in May 1938, they had stiffer wings and a maximum diving speed of 470 mph indicated airspeed, nearly 100 mph faster than that of the prototype Spit. Jeffrey Quill and the other test pilots in the programme immediately found that, even at 380 mph, the controls were unsatisfactorily heavy. A lengthy programme of trial-and-error fixes ensued. Quill: "It may seem strange that no one at that time suggested that severe tactical limitations would be imposed by the heavy ailerons. This can probably be accounted for by the fact that the Air Staff regarded the Spitfire (and the Hurricane) as being primarily for home defence and hence bomber destroyers rather than as being designed for dog-fighting against other fighters. The air threat to this country was seen to comprise massed raids by enemy bombers and it was not thought that German single-seat single-engined fighters could ever appear over Great Britain because they had no bases within range. Moreover those surviving officers with operational fighter experience from the First World War had fought in Sopwith Pups, Camels and SE5As in fighter-to-fighter dog-fights over the static Western front, and their aeroplanes had been capable of not much more than 100-120 mph. Now here was a fighter flying more than four times as fast and the concept of a dog-fight at such speed seemed to them grotesque. Contemporary wisdom, certainly amongst senior officers, therefore simply assumed that the day of the dog-fight was long past.

"I started on this programme on Spitfire X4268 within two days of returning from Hornchurch, and we

SERGEANT PILOTS OF A SPITFIRE
SQUADRON AT HAWKINGE
BETWEEN AIR ACTIONS IN THE
BATTLE OF BRITAIN.

worked our way through a whole series of variations including geared tab balances and blunt noses, but nothing showed much result, while some, indeed, made matters worse. George Pickering and Alex Henshaw flew on this programme too. Then, just as we were beginning to feel thoroughly depressed about the situation, I tested, on November 7th, a pair of metal-skinned ailerons with thin trailing edges on Spitfire R6718. I remember doubting, as I taxied out, whether they would make much difference. In fact they changed the situation in the most spectacular way, being much lighter at the top end of the speed range without any loss of effectiveness at low or medium speeds. The aeroplane was transformed. I flew it four times more that day and threw it about violently to try to find some snag or some undesirable feature, but I could not. George and Alex flew it and agreed it was a huge improvement."

Reactions from the test pilots at Farnborough concurred, "excellent control", "consider all Spitfires should be modified as soon as possible".

RAF Squadron Leader Willie Wilson worked with Allen Wheeler in the experimental flying department at Farnborough. In late November 1940 he was briefly attached to 74 Squadron at Biggin Hill, Kent, to go with its pilots on defensive patrols and evaluate the Spitfire in action. His conclusions: "At 30,000 feet the Spitfire is very sluggish and unmanoeuvrable and the petrol gauges generally freeze up, which is very dangerous in an aircraft with a short endurance. Owing to its engine the Me 109 can be banged into a dive, whereas with the Merlin the slightest negative G causes

the engine to die. Another point which worries the pilots is that they do not carry a reserve supply of oxygen fitted to the parachute for baling out at high altitudes. During . . . high-speed dives following Messerschmitts down from over 30,000 feet the windscreen and hood mist and ice up just when it is essential to have a good view of the enemy. Another factor causing concern to the pilots in a dive is the inability to open the hood in emergency; it is of very great importance that a jettisoning device be incorporated.

"In view of all of the above it can hardly be said that the RAF Spitfire squadrons have superior fighting material to the enemy, and it is my considered opinion that the only reason why we are just managing to maintain the balance of fighter power is the outstanding flying and leadership of the pilots. If these same pilots are given equipment even only as good as the Germans they would be able to beat down everything that comes within sight." Wilson submitted a list of suggestions, nearly all of which were implemented within the next year: 1. Immediately fit Merlin XXs and at the same time develop an engine which enables a squadron to fly in formation at 40,000 feet, 2. Fit at least two cannons to every aircraft, 3. Make the ailerons metal covered in order to increase manoeuvrability, 4. Arrange for the hood to jettison as on the 109, 5. Fit a small oxygen reservoir to the parachute, 6. Issue the squadrons with Gnomist to prevent misting and icing up, 7. Devise a non-freezing up petrol gauge, 8. Fit VHF instead of the TR9B radio, 9. Instead of using a bulky life-saving jacket filled with kapok, have an inflatable air jacket built in many sec-

tions so that punctures due to bullets do not cause a complete collapse, 10. Improve the clocks in the contactors so that they can be set on the ground, 11. Adopt the French type of harness lock which does not protrude and get caught up, 12. A non-bulky 'Sidcot' suit should be designed preferably incorporating a lifejacket, 13. The new-type helmet, mask and goggles should immediately be released to the fighter squadrons giving more freedom of head movement, 14. An endeavour should be made to make our engines work during negative G."

The radio and television commentator Raymond Baxter was a Spitfire pilot in WW2. His favourite type was the Mk XVI clipped-wing Spit, which carried up to 1,000 lb of bombs. He flew it with 602 Squadron. "I got enormous satisfaction from dropping that little lot onto the enemy, invariably on pin-point targets. The most proud moments of my entire life came on the very few occasions—and those by chance—when I led thirty-six fully-armed Spitfires into action. A mere matter of months since I had sworn my allegiance to my Sovereign Lord, King George VI, his heirs and successors, I met my first Spitfire. It was not love at first sight. I'd already been in love with it for ages. My earliest passions and interests—engineering, power, speed and flight—were here encapsulated in a single beautiful machine and given to me, not just to play with, but to put to the purpose for which it had been created. But now we had come to the moment of truth."

Baxter remembered one particular Spitfire operation: "One evening in early March 1945 Squadron Leader Max Sutherland DFC, CO 602 (City of

FELDWEBEL OSCAR BOESCH AND HIS FOCKE WULF FW190 FIGHTER.

Glasgow) Squadron, gathered his four senior pilots to the bar at RAF Ludham. After the first round of drinks, his brown eyes blazing with challenge, he said 'Listen chaps, I've got an idea.' Our minds became concentrated; we knew Max to be unpredictable, to say the least.

"At that time we were flying our socks off in Operation Big Ben—the anti-V2 campaign. On several occasions during those ops we saw the white trail of V2 rockets arcing up at incredible speed and height towards their indiscriminate targets. It made us the more determined to strike back as hard as we could. Equipped with Spit 16s we were dive-bombing every reported or suspected launch site, and dive-bombing, skip-bombing and strafing interdiction targets throughout occupied Holland. In addition we were committed to normal fighter duties, readiness, bomber escort (U.S. & RAF), air-sea rescue,

and shipping and reconnaissance patrols. Four sorties per pilot per day were not uncommon.

" 'Just outside The Hague,' said Max, 'is the former HQ of Shell-Mex. It is now the HQ of V1 and V2 operations. I have worked out that the width of the building equals the total wingspan of five Spitfires in close formation.' He paused to let it sink in. 'I reckon we could take it out.' Then he said, 'What do you reckon, Bax?' I took a long slow draught from my Guiness and said, 'Might be a bit dodgy, Boss.'

"The concentration of heavy and light flak from Den Helder to the Scheldt was our daily experience, and later we learned that no less than 200 batteries of well-manned guns would lie in our path. 'Yeah, I know,' said Max. 'But we'll get 453 to lay on a diversion and we'll go in flat and low.' 453 were the Australian squadron in our Wing, led by Squadron Leader Ernie Esau, another

major character. They were our neighbours on the airfield, with whom we had developed a close bond both in the air and at the pub.

"So it was agreed and, somewhat surprisingly, approved by Group. After a disappointing abort, because cloud obscured the target on 18th March, the attack was delivered precisely according to plan. We peeled off from 453 at about 8,000 feet, crossing the Dutch coast, and Maxy transmitted the seldom-used codeword, 'Buster' (full throttle), and 'Close Up'. He then led us in a perfectly judged diving arc, in which the controls became increasingly heavy and, therefore, more demanding for close formation. We flattened out at about 100 feet with the target dead ahead and square in our gyro-sights—range about 300 yards. We let go with our two x 20 mm cannon and 0.5 in machine-guns and released our 1 x 500 lb and 2 x 250 lb eleven-second delay bombs 'in

our own time'. Then as I cleared the roof of the building, I looked ahead. And approaching me at eye-level, and near enough 400 mph, was this black cockerel atop the weather vane on the church spire across the road!

"The PRU photograph, which we had studied, showed the church (which we were determined not to hit), but not the height of its spire. With no space to turn, I could only tweak the stick and say, in my head, perhaps two seconds later, 'Thank you, God.'

"Then the Boss damn near got his tail shot off. Our pre-attack briefing had been that the moment we cleared the target we should fan out, continuing at roof-top height. So five Spitfires swept at extremely low level across the centre of the Dutch capital. This, as was hoped, presented the anti-aircraft gunners with such a variety of fast-moving targets that the amount of flak we encountered was comparatively light. Unfortunately, however, Max Sutherland—as he had done on many occasions previously— decided to pull up to have a look back and assess the extent of the damage which we had inflicted with our eleven-second delay bombs. This, of course, attracted the concentration of anti-aircraft fire onto him. Nevertheless, he had the satisfaction of observing that the whole of the Shell-Mex building was occluded by a cloud of smoke, flame and dust.

"But, almost immediately, Max called us again, saying 'Tarbrush (I think that was the call-sign), proceed as planned to rearm and refuel. I have been hit. Bax, please come and have a look at the damage.'

"So, clear of the built-up area where the flak was, Max pulled up to about 3,000 feet and, as the rest of the formation sped towards their refuelling point, I closed to him and had a careful look at his tail. 'Boss,' I said, 'you've got a big hit on your starboard elevator.' 'Is it still flyable, do you think, Bax?' he said. 'Up to you, Boss,' I said, 'but I'll keep a careful eye.'

" 'I will continue at bale-out height,' he replied. 'Please stay close.' So I did that, at about 3,000 feet and on his starboard side, slightly behind and below, keeping a very careful eye on that severely damaged tail. We made it to the circuit—at Ursahl, our planned refuelling point near Ghent in Belgium— and as Max had his airspeed indicator and flaps working, I remained close and watched him make a perfect landing. Then I followed him in and we had time to take a look at what we had achieved by comparing notes with our mates, who were all safely down and undamaged.

"Max got a bar to his DFC for that, and the rest of us got a Collective Mention. The BBC Nine o'clock News that night reported, 'During the day, RAF fighter bombers continued their attacks on selected ground targets in occupied Holland.' Just so!"

The targets for the bombers of the American Eighth Air Force on October 13th 1943 were the ball-bearing factories of Schweinfurt, south-east of Bremen in central Germany. The immense complex of plants was within four months of producing most of the bearings for the aero engines of the Luftwaffe. What were predicted to be the last favourable weather conditions of the year afforded the bombing planners an unparalleled opportunity to eliminate that production in a single strike. But the logistics of the attack had to be organised and implemented in only forty-eight hours and in total secrecy. On nearly 100 air-fields around Britain, hundreds of heavy bombers and 1,300 British and American fighters had to be made ready for the raid. The B-17 Flying Fortress bombers would be over enemy territory for more than four hours and their crews were briefed to expect a formidable reception from the Luftwaffe fighter force. Upwards of 3,000 Messerschmitts and Focke-Wulfs, on bases from Denmark to Belgium, would be waiting to meet them. The Americans needed all the fighter protection they could muster and they requested the help of RAF Fighter Command which, for the first time in the war, provided Spitfires for long-distance bomber escort work over Germany. Having been designed as high-speed fighter-interceptors, Spitfires were not prepared for the radius of action required in raids like Schweinfurt. They needed the installa-tion of 90-gallon auxiliary fuel tanks, 800 of which had to be produced in less than three days for fitting to the bellies of the Spits. These, and hun-dreds of other details, had to be dealt with before the massive attack could be launched. And then, at the eleventh hour, the raid was put back to the early hours of October 14th.

The plan called for sixty B-24 Liberator bombers to participate in the mission, but poor weather conditions resulted in only twenty-nine of the planes joining the party. The B-24s were reassigned to a diversionary strike towards Emden. The 320 B-17s dis-patched on the attack had thinned to 229 by the time they reached the tar-get area, owing to the weather, mechanical problems and fierce enemy opposition. The take-offs had begun at eight a.m. Nineteen

squadrons of American P-47 Thunderbolts took off at 10:40 a.m. to take over from the RAF Spitfires which had left at 9:00 a.m. on their escort duties. Ten Spitfire squadrons were allocated for the raid and were gathered on four airfields in Norfolk. The pilots had no experience of flying such aircraft with the extra weight of their auxiliary belly fuel tanks and two of them crashed. Tyres burst and there were a number of air-locks with the belly tanks. One of the Spitfire pilots, Pierre Clostermann, recalled the expe-rience: "Jacques [Remlinger] and I were among the victims. Landing on our flimsy tyres, with ninety gallons of juice under the belly and 150 more in the wings and fuselage was tricky—like landing on eggs, as Jacques put it. Seething with rage, we watched the swarm of Spitfires disappear towards Germany in the morning mist."

While the fitters and riggers went to work on the remaining Spitfires, preparing them for their second mis-sion of the day, Jacques and Pierre went to sleep under their aircraft. The Spitfire squadrons returned from escorting the bombers at 11:45 and were immediately refuelled and rearmed. The tired pilots tried to relax briefly with sandwiches, tea and ciga-rettes but there was little conversation. The op had produced no enemy fight-ers as of 10:30 a.m. when the Spitfires had had to leave the bombers.

At Bradwell Bay, Jacques, Pierre and the other pilots of 132, 411, 453 and 602 Squadrons began taking off again at 12:04 p.m. The two Frenchmen flew as number three and four respectively to Max Sutherland in the lead of Yellow Section. Pierre: "The weather, which had been so fine till midday, had deteriorated and big banks of

cloud and mist rose vertically from the ground like ramparts. Going through one of these big cumulus Jacques and I had lost contact with the rest of the squadron. Now we were lost in the inferno and we stuck frantically together, trying to get to the ren-dezvous [with the bombers]. Somewhere on the right below the mist must be Emden and the rich canal-bordered pastures of North Holland. Far behind us already, the Zuider Zee. Up in the air it was a nightmare. I had never seen anything like it. Clusters of flak appeared from the void and silently hung on the flanks of the clouds. Space brought forth swarms of German fighters—a disquieting example of spontaneous generation."

For the first time in the year that the Eighth Air Force had been bombing Nazi targets in occupied Europe, the Ju 88s and Me 410s of the German Air Force were successfully scattering and breaking up the precise box defensive formations of the American bombers. It was a scene of chaos and near panic as the B-17 crews struggled to protect themselves against the rockets and cannon shells and the desperate atten-tions of the Focke-Wulf 190s running in for the kill. Pierre: "It was a miracle that we had not yet been brought down! Twisting and turning, firing off our guns, we had succeeded in gain-ing quite a lot of height over the main scrap. I had exhausted half my ammu-nition. Suddenly Jacques spotted in the middle of the sky dotted with para-chutes and burning aircraft about forty Focke-Wulfs pouncing on four Fortresses which were lagging behind to protect a Liberator, one of whose engines was in flames. What could we do?—it was impossible to call for help

A EUROPEAN STAR
WITH BATTLE OF
BRITAIN CLASP;
ABOVE GERMAN
AIRMEN IN 1940.

in this infernal scrum. All the Spitfires, as far as the eyes could reach, were whirling in dog-fights.

" 'Attacking!'

"Sights switched on, finger on firing button, together we rolled on our backs and dived on the Focke-Wulfs milling round the bombers. They were attacking from every side—front, side and rear. One of the Fortresses went into a spin, slowly. Another suddenly exploded like a gigantic flak shell, and the explosion tore a wing off the one to its right. A big dark mushroom spread out, incandescent debris dropping from it. The now asymmetrical outline of the Fortress grew smaller and fainter, falling like a dead leaf. Like shining new nails on a wall, one, two, four, six parachutes suddenly dotted the sky. I tightened my stomach muscles, put my feet on the top pedal of the rudder bar to resist the centrifugal force, swallowed hard to get the bitter taste out of my mouth and pulled out violently. Before I had time to register, my finger had instinctively pressed the firing button. A burst at the Focke-Wulf, who for one second filled my windshield. Missed him! Surprised, he stalled and fell away. Jacques fired on him and missed him too—but a grey Messerschmitt, its wings edged with fire, was after him. I yelled: 'Look out, Jacques! Break right!' Quickly, I put all my weight on the controls, the ground whipped round—but too late, the Messerschmitt was out of range. I was drenched in sweat.

"In front of me two Focke-Wulfs were converging to attack a Fortress drifting like a wreck. A glance at the mirror: Jacques was there. The red filaments of my sight encircled a green and yellow Focke-Wulf—Jesus, how close he was! The wings of my Spit

shuddered from the hammer-blows of my two cannon . . . three flashes, a belch of flame and a grey trail unfurled in his wake! Then I saw a sheet of flame on the flank of a cloud, just where Jacques' aircraft was at that moment—my heart missed a beat—but it was his triumphant voice shouting in the radio: 'Did you see that, Pierre? I got him!'

"Thank God, it was a Focke-Wulf, and out of the corner of my eye I saw his Spit swaying fifty yards away. What a relief! Suddenly a thunder clap, a burning slap in the face. My eardrums were pierced by the shriek of air through a hole just torn by a shell through my windshield. Bang! Another . . .

"For a time I lost all notion of what was going on. For ten minutes I blindly followed Jacques's instruction over the radio; when I picked up the thread again we were in the middle of the North Sea. On my right was a Fortress, holed like a sieve but flying all the same.

"England at last. Just inland I could make out four crashed Fortresses in the fields. We landed at Manston after the Fortress, exhausted, drained."

First Lieutenant Raymond W. Wild, of the 92nd Bomb Group (H), 8AF, piloted the lead B-17 in the first American raid on Berlin. Between October 4th 1943 and June 2nd 1944, he flew a then-standard tour of thirty missions over German-held territory in continental Europe. His third mission was the Schweinfurt raid of October 14th. Wild had been keeping a diary beginning with his first mission. After the Schweinfurt raid he put the diary away having concluded that no one was likely to survive the tour.

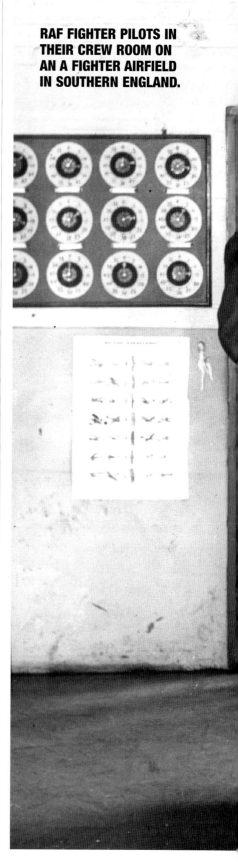

RAF FIGHTER PILOTS IN THEIR CREW ROOM ON AN A FIGHTER AIRFIELD IN SOUTHERN ENGLAND.

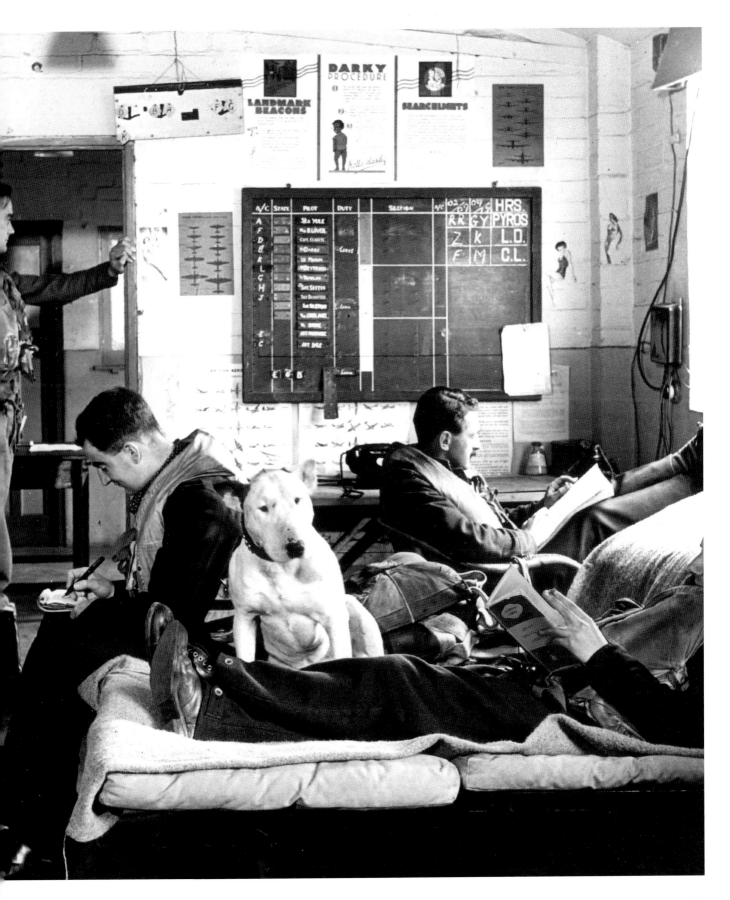

Wild: "I was a devout coward. But what happens is, pride makes you get into that airplane, and pride makes you stay there and keep going when what you really want to do is to turn back while you still can. They had that tradition that an Eighth Air Force sortie never turned back from a target. So, hell!, you didn't dare turn back!

"I remember that just before my first raid—the one where you are really frightened to death—I went into the john in the operations tower. Didn't have to go, but just went in and sat on the john. That was when the song *Paper Doll* had just come out and somebody had written all the words on the wall. Well, just through nothing but being nervous I sat there and memorised those words." Wild's Flying Fortress returned safely to his Podington, Northamptonshire, base from the mission that day. "From then on I sat in that same john every mission morning. And you know, to this day I know every word of *Paper Doll*."

Then came Schweinfurt. Wild: "The 88s which were, I think, mostly used for flak, were tremendously accurate, just fabulous. They used two types: pre-determined and barrage. In barrage, there'd be a flock of guns and they'd shoot at one spot in the sky and keep shooting at it. The other was pre-determined aiming at planes. The most frightening one was the indeterminate shooting at a spot in the sky. You had to go through that spot when they weren't shooting. Emden, Kiel, Wilhelmshaven, Munich, Berlin . . . I think they did both kinds at all of those targets. But Schweinfurt was murder. I'm sure they shot barrage because they had so damned many guns. The German fighters stayed pretty much out of the flak, but on

Schweinfurt they did come through the flak. It's one of the few times they flew in their own flak, but they were probably under orders. That indeterminate flak that was coming up, there was nothing you could do about it. This was something that was gonna happen. It was impersonal as hell."

The Schweinfurt factories had been dealt a heavy blow. 197 German fighters had been shot down, but the American Eighth had suffered too. Of the participating B-17s, only about fifty were still airworthy. More than 100 Allied crews had been killed or taken prisoner. Fifty-one Allied escort fighters, including many Spitfires, were downed.

Brian Kingcome was one of the outstanding British fighter pilots of the Second World War. He joined the RAF in 1938 with a permanent commis-

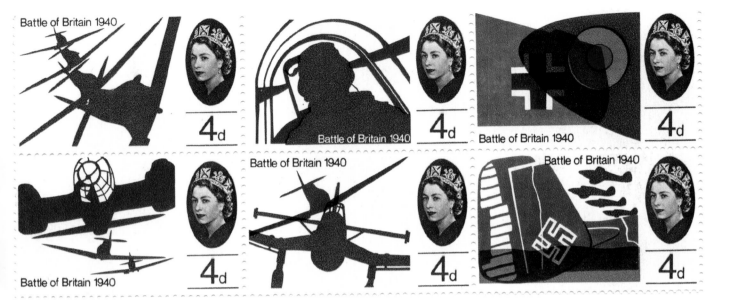

LEFT: RAF FIGHTER PILOTS RELAXING IN THE SUN BETWEEN SORTIES IN 1940; ABOVE: ROYAL MAIL BATTLE OF BRITAIN COMMEMORATIVE STAMPS ISSUED IN THE 1980S.

sion via the cadet college, Cranwell. "I always regarded 92 Squadron as my personal property. I led it through, what was to me, the most exhilarating and treacherous part of the war, the Battle of Britain at Biggin Hill. I gained and lost many good friends, and in front line operations I was with 92 longer than any other squadron." Of the Spitfire he offered: "The first time I flew the Mark IX I could hardly believe the experience. The effect was magical. I had expected an increase in power, but nothing to match the reality. To enhance the dramatic effect, the second stage cut in automatically without warning. One minute there I was, relaxed and peaceful, as I climbed at a leisurely pace towards 15,000 feet, anticipating a small surge of extra power as I hit the magic number. The next minute it was as though a giant hand

had grabbed hold of me, cradled me in its palm like a shot-putter his weight and given me the most terrific shove forwards and upwards. The shock was so great that I almost baled out. It literally took my breath away. It was exhilarating, a feeling I could never forget. I yearned at once for a chance to demonstrate this astonishing new tool to the Germans."

After the Battle of Britain, Spitfires took the offensive in fighter sweeps over occupied Europe, including large-scale operations known as Circuses, involving up to eighteen squadrons. They also took an active part in escorting Allied bombers on their raids on Nazi targets. Taking their Spitfires with them, the American pilots of the three RAF Eagle Squadrons assembled at RAF Debden, in Essex, on September

29th 1942, to transfer officially to VIII Fighter Command of the USAAF. As part of the growing Eighth Air Force, the resulting Fourth Fighter Group would quickly become prominent in the role of shepherding the American B-17s and B-24s in their campaign of daylight precision bombing of German-occupied targets initially, and later of Germany itself. They would ultimately trade their Spits for P-47 Thunderbolts until the arrival of their P-51 Mustang long-range escort fighters early in 1944. Mk IX Spitfires with added fuel tankage were soon accompanying the bomber streams much further into enemy territory. The bigger and more powerful Rolls-Royce Griffon engine entered front-line service in 1942 with the conversion of 100 Mk XII Spitfires which had been

AN RAF POLISH SQUADRON SPITFIRE PILOT; RIGHT: VIC BUONO AND ART ROSCOE, OF AN AMERICAN EAGLE SQUADRON.

built as Mk Vs, Mk VIIIs or Mk IXs. The new Griffon planes had been specifically designed for low-level operation against marauding enemy fighter-bombers on England's south coast.

In another important role, the Griffon-powered Mk XIV Spit, with its distinctive five-blade Rotol propeller, was a very effective counter to the German V-1 flying bomb, which was also known as the Buzz Bomb or Doodlebug. With their exceptional speed, the Mk XIVs operated against the V-1s in flight as well as attacking their launch sites in the Pas de Calais. Continuing development of later Spitfire marks enabled the type to excel in additional roles, not least being essential reconnaissance in the various theatres of the war.

Spitfires were key in the defence of the Mediterranean island of Malta

and in the campaign against the German and Italian resupply and reinforcement effort from March 1942. They were delivered to the area by the aircraft carriers HMS *Eagle*, HMS *Furious* and USS *Wasp*. Together with Hurricanes, the Spits were particularly effective against the enemy forces.

Fitted with tropical filters, Spitfire VBs were operational with their Hurricane sisters in North Africa and the Middle East from the spring of 1942, disrupting German air reconnaissance and harassing the enemy with their presence.

The Seafire, a naval version of the Spit, was initially a conversion of the Mk VB airframe, intended for low and medium altitude air combat and reconnaissance. These aircraft saw considerable action in the Mediterranean in 1943 and the Pacific in 1945. It

was in the North African war theatre that the Seafire made its combat debut. The Seafire played an important part in the Allied invasion of Sicily on July 9th 1943, providing vital air cover over the beachheads.

Spitfire units operated in Greece and Yugoslavia and later in Italy. In the south Pacific, Spitfire Vcs and VIIIs proved their worth in the hands of the Australians against the Japanese. They also scored brilliantly in the Burma-India theatre where one of their main assignments was the protection of transport aircraft bringing in supplies to besieged soldiers at Imphal, the point where the Japanese were attempting a major breakthrough to India.

The Spiteful was the last development in the Spitfire line, coming near the end of the war, and it looked less like the early Spits than

any of its predecessors. It was designed primarily to prove the application of a laminar flow wing on a high-speed piston-engine fighter. Spiteful was less successful than Vickers-Supermarine had anticipated and was deemed unsuitable for squadron service. A planned naval version of Spiteful—the Seafang—also failed to demonstrate sufficient improvement over earlier Seafires to justify quantity production. The last of the test Seafangs had Griffon 89 engines and two three-bladed Rotol propellers. They were, in fact, judged acceptable for carrier operations in early 1946, but by that time the Admiralty had opted for another Supermarine aircraft instead—the first Royal Navy jet fighter, the Attacker.

After the war some Spitfires that were surplus to RAF requirements went to the air forces of many other nations, including Greece, Czechoslovakia, France, Belgium, Denmark, Norway, Sweden, The Netherlands, Ireland, Italy, Portugal, Yugoslavia, Syria, Turkey, the USSR, Israel, India, Burma and Thailand. Ironically, in the Arab-Israeli Wars of 1948-49, Spitfires went into battle against Spitfires.

"You will rarely hear an RAF pilot speak ill of his aircraft. There are two reasons. Firstly, it is a matter of professional pride, and, secondly, one of confidence. Every experienced airman knows that an aeroplane which is unloved is far more likely to kill its pilot than one whose characteristics and temperament are treated with the consideration and tolerance of a lover towards his mistress, however trying that relationship may sometimes be."

AMONG THE MOST IMPORTANT ROLES OF THE RAF SPITFIRE IN WWII WAS AS ESCORT FOR THE AMERICAN HEAVY BOMBERS ON THEIR DAYLIGHT RAIDS TO GERMAN TARGETS.

THE SUPERMARINE WALRUS SEA-
PLANE PLAYED A VITAL PART IN
RESCUING RAF AIRMEN WHO HAD
BEEN BROUGHT DOWN IN THE
ENGLISH CHANNEL DURING WWII.
IN THE BACKGROUND AN RAF
LYSANDER ACTS AS A SPOTTER
AIRCRAFT IN THE RESCUE EFFORT.

TEST PILOT

Supermarine Chief Test Pilot Jeffrey Quill recalled that when Alex Henshaw first arrived at Eastleigh to join Quill and George Pickering in the flight testing work, Henshaw quickly familiarised himself with the Spitfire and the test schedule. Quill: "Our technique in bad weather was to take off from Eastleigh, creep along the railway line towards Fareham at 'nought feet', cross the Solent and fly to a point south of St Catherine's Point to begin our climb. We would then let down and return by the same route. Before long, Alex was doing this in weather which was as bad as it was possible for anyone to fly in without 'aids'. On one occasion, when the solid base of wet and rain-laden stratus cloud was very low indeed, I had just landed back at Eastleigh and was wondering whether it was not pushing my luck to make another sortie. Then Alex's Spitfire appeared through the murk; as he crossed the airfield, he performed a slow roll as close to the ground as I had ever seen it done in that category of aeroplane. His wing-tips literally brushed the base of the cloud. Because of the rain there was no horizon, and there was absolutely no margin for error either way. The manoeuvre was perfectly executed, yet it shook me considerably. He came into the flight office, and it was as we were stripping off our Mae Wests I said, 'You're asking for it, Alex, aren't you?' This developed into a sharpish row between us, which died down as quickly as it flared up. It was already obvious to me that in Alex Henshaw we had a pilot of the most exceptional skill and ability and totally unorthodox in his approach to flying. Things that would be regarded as lunacy by normal standards seemed perfectly manageable to him. He seemed to be a sort of aeronautical phenomenon, and there was no point in trying to impose restrictions upon him which he would not accept anyway.

"For some months I watched Alex's flying with growing amazement and respect. There was a period when I thought it was only a matter of time before he 'bought it' but thankfully he never did. Later in 1940 Alex moved up to Castle Bromwich to take over the mammoth production testing task there."

When Alex Henshaw first went to Eastleigh he liked what he saw. He liked the people there and he appreciated the smaller scale and more comfortable atmosphere than he had known at Weybridge where he had recently been working for Mutt Summers. "I was out early next morning and was the first to arrive at our little office on Eastleigh aerodrome. I had been thinking about flying my first Spitfire most of the night and sincerely hoped that I should not make a hash of it. As I had nothing else to do, I strolled the hundred yards or so along the final assembly block. Arthur Nelson, whom I had met the day before, was in charge and I asked him if he would mind if I sat in a fully assembled Spitfire to familiarise myself with the instruments and controls and in particular the rather clumsy chassis retraction gear. He replied, 'I'll do better than that; you can nip into this machine on the test-rig. There you can operate the flaps and lower and lift the wheels to your heart's content.' I was very glad of this offer. Although the cockpit was quite simple and straightforward, the chassis hydraulic pump handle was the same as one would see in the local garage to operate small jacks, except that the control handle in the Spitfire was big and solid and rather reminded me of the tiller on some sailing boats that did not use a wheel. It required quite an effort to operate the pump and I saw immediately that you had to change hands from the throttle and the control column to pump after take-off. Nelson put me on my guard when he laughingly said, 'Bet you a pound you do a corkscrew the first time you go up.' "

Henshaw was offered the position of Chief Pilot at Castle Bromwich, but turned it down. He did not like the idea of working for "a vast sausage-machine turning out aircraft". He much preferred the mixed type of flying he was enjoying at Southampton (Eastleigh); he worked well with Jeff Quill and George Pickering. But after he and his wife Barbara had made about a dozen trips to Birmingham and the new factory, they realised how much better they both seemed to feel and how much more energy they seemed to have up in the Midlands. In the Southampton area they had to force themselves to play any sport, and gardening was always an effort. Their limited spare time was spent in a perpetual state of exhaustion. So after further consideration, and Barbara's admission that she would actually prefer to live in the Midlands, Alex went back and asked if he could withdraw his refusal.

The first Spitfire mark produced at the Castle Bromwich shadow factory was the Mk II. Alex remembered that

it featured many improvements on the Mk I, the most important being the elimination of the large pump-handle and the fitting of a power-operated control for the undercarriage. It also carried a constant-speed propeller and the Merlin XII engine which gave the aeroplane a slightly improved take-off and climb.

There was a war on, as if anyone needed reminding. "Every day we expected the invasion as the bombing got worse day and night. We fared better in the day than Supermarines, but at night I felt sure it could not go on for much longer without something cracking. The Germans seemed to find the factory at Castle Bromwich almost every time and one got hardened to the sight of a direct hit on, say, the machine shop before all employees had gone into the shelters; as day slowly dawned it was common to find bits of bodies mingled with valuable equipment blasted to small pieces and hanging gruesomely from what had been the roof. The spirit of the British people at that time was magnificent and quite irrepressible and within days there would be a temporary roof on, new machines would be moved in and the shop would be back into production. One of the worst weapons of that summer was the delayed action bomb. These were scattered all over the place, particularly on the aerodrome and around the factory, and it was a little disconcerting when taking off for one to suddenly explode nearby.

"Our losses in aircraft during training periods were extremely high, particularly with the Spitfire. Alex Dunbar said to me one day that he had been talking to the Chief of Fighter Command and had thought

that if I went round to some of the Operational Training Units and demonstrated the Spitfire it would do a great deal of good. I didn't enthuse over the idea but said if I received a request from an OTU I would of course be pleased to go. The first visit was to Hawarden, which had changed into one of the most important OTUs in the country for final Spitfire training. Here as well as hundreds of novice trainees were some of the best pilots in the Air Force; I was fully aware that whilst I might have got away creditably with some wild aerial gyrations in front of an ill-informed group, this time my audience would be critical. In fact some might welcome the opportunity to pick holes in a young amateur.

"I was welcomed at Hawarden, courteously. After a short chat about the object of the exercise, I took off somewhat expectantly as the Commanding Officer said he had not bothered to stop all the flying: he expected I would only be a few minutes. I flew my best without taking undue risks and not cutting my margins too fine. I think I felt at the time that my show might have been all right but could have been better as part of my concentration was broken looking out for stray aircraft. When I landed any apprehension I may have felt was quickly dispelled as I glanced at the faces of the senior officers surrounded by numerous young enthusiastic OTU pilots all surging forward to greet me. Wing Commander Donaldson was the first to speak: 'I've seen them all, but if I had not been here today I would never have believed a Spitfire could fly like that,' he said. Naturally I was relieved and very satisfied, but as it was then coming on to rain I made my excuses and had a wet, uncomfortable flight back to Castle Bromwich, where Barbara waited for me patiently with our lunch ruined."

Wing Commander Peter Ayerst was one of Alex Henshaw's junior officers at Castle Bromwich and gave a graphic description of Henshaw at the factory airfield. "He was professional in everything he did, whether it was flying, writing reports or briefing the pilots on specific items. The professionalism was particularly visible in his flying displays. To those who were lucky enough to see Henshaw flying a Spitfire—it was astounding. The execution of his manoeuvres was crisp and precise and his performance breathtaking. Those who haven't seen Henshaw's displays would never believe that a Spitfire, or for that matter any aircraft, would be able to withstand such demonstrations; you had to see it to believe it. A continual stream of high-ranking visitors including Winston Churchill, HM King George VI and a long list of foreign diplomats came to see the massive production capabilities of Castle Bromwich. They were treated to Henshaw's flying display as a matter of routine. One move that Alex had perfected: a bunt (outside loop) at about 1,500 feet, was a manoeuvre that was extremely dangerous at such a low altitude. With the throttle pulled back, the nose pulled up, it had the effect of stalling the Spitfire. The nose would suddenly drop towards the ground and Henshaw would then be flying inverted at house height adjacent to the spectators, pulling out of it. There were

other tricks—diving flat out, then pulling into a vertical roll and then at the top he would stall turn, drop off and come screaming down again. It was truly fantastic. A consummate aviator. No one, including the Chief Test Pilot for Supermarine, Jeffrey Quill, could match his sheer breathtaking performance and flying prowess in a Spitfire."

Sometimes bad things do happen when flying high-performance aircraft, especially in wartime. Henshaw: "I had just had a forced landing, which had shaken me although I was completely untouched. During a take-off westwards over the factory the engine cut just as I was over the boundary of the airfield. If I had followed the unbreakable rule never to turn down wind on

ALEX HENSHAW, CHIEF TEST PILOT FOR VICKERS AT CASTLE BROMWICH IN WWII.

engine failure but keep straight on, I should have landed on top of Dunlops or our own factory, so I pushed the nose down quickly as I turned back, losing height so rapidly that if the old Castle Bromwich clubhouse roof had been a foot higher I should not have made it. I flicked the emergency landing gear lever downwards but the margins were so fine, and it had happened with such suddenness, with the machine almost dropping onto the ground downwind, that only one leg of the chassis locked into position. As the machine rocketed over the grass, I felt the port leg start to fold up and the wing first scraped and then dug into the soft ground; spinning round we jerked to an abrupt halt. I suppose to those watching it was just a neat piece of flying, other than the chassis collapsing, but only I knew how close it had been before I appeared stepping calmly from the cockpit; I paused to offer a silent prayer of thanks. During the investigation a large split-pin was found in the magneto distributor housing; this could only have been sabotage at the point of assembly, as no such type of split-pins were used on the whole magneto, and before that cover was bolted up it must have been subject to a very close and careful examination. We had heard all sorts of rumours about incidents with torches and rockets at night and also sabotage to certain engines being found in flight, but I had not believed them and thought the stories exaggerated. I knew most of my own men extremely well by now, but of course we were taking on increasing numbers and these were complete strangers."

Most people who have been associated with the Spitfire have a favourite

mark, including Alex Henshaw. "Early in 1941 production at Castle Bromwich changed over to the Spitfire Mark V. In appearance it was almost identical with the Mark I or Mark II when fitted with the 'A' type wing. When fitted with the 'B' wing it had two 20 mm cannon and four .303 machine-guns; with the 'C' wing four 20 mm cannons only; and for the 'D' wing there was the combination of two 20 mm cannons and two .5 machine-guns. The empennage and nose had not changed in shape except on the tropicalised version, which had a snub-nosed appearance due to the large intake filter which spoiled its otherwise clean profile. The camouflage paint scheme had also not altered, except again, on the tropicalised version—they were painted a light brown and sandy yellow which after the drabness, to which we had now long been accustomed, looked to our war-weary eyes almost festive. With the various modifications that had been creeping in, such as improved armour-plating, self-sealing tanks and heavier guns and ammunition, etc., the gross weight had now reached almost 1,000 lb more than the first Spitfire I had flown at Eastleigh. We were fitting the Merlin 45 which gave another 10-12 mph in maximum speed at 20,000 feet, and we could now reach 25,000 feet in the same time that the Mark I took to reach slightly under 20,000 feet. The Mark V came out in several guises but the one I enjoyed most of all was the model fitted with the Merlin 50M engine, and the 'A' type wing. This was a cropped blower or de-rated engine, designed to give its maximum power at 5,500 feet. It had, of course, a big advantage at low level and with

the enormous increase in boost pressure, it was a joy to fly. In fact it was the only Spitfire during a demonstration that I felt able to take-off, lift the wheels up, pause and then pull up firmly but smoothly into a vertical loop with a slow roll off the top to finish over the centre of the airfield. If I had to make a choice of all the numerous marks of Spitfires—and there were over thirty-six of them—this is the one I would have picked for a low-level display."

The brilliance and singularity of Jeffrey Quill and Alex Henshaw, the greatest of the wartime British test pilots, is wonderfully reflected in this comment by Henshaw, ". . . whilst we knew very little by today's standards, there were many very obvious ways overstressing of engine or airframe could be checked. Those who think a test pilot's job is to test an aircraft to near destruction probably have been watching Hollywood films: this of course is sheer nonsense, particularly in wartime production testing. Here the job of the pilot is to test within design limitations and for a specific purpose: to prove that each completed aircraft conforms to the previously approved and tested design and is 100 per cent sound and suitable to be handed over for active service. If a production aircraft were overstressed during the process, it would, I think, be a very serious reflection upon the test pilot concerned."

Of all the test pilots who worked for Alex Henshaw at Castle Bromwich, one stands out in his memory. "Venda Jicha was the best Spitfire pilot I ever had at Castle Bromwich. He had been in the Czechoslovakian Air Force and

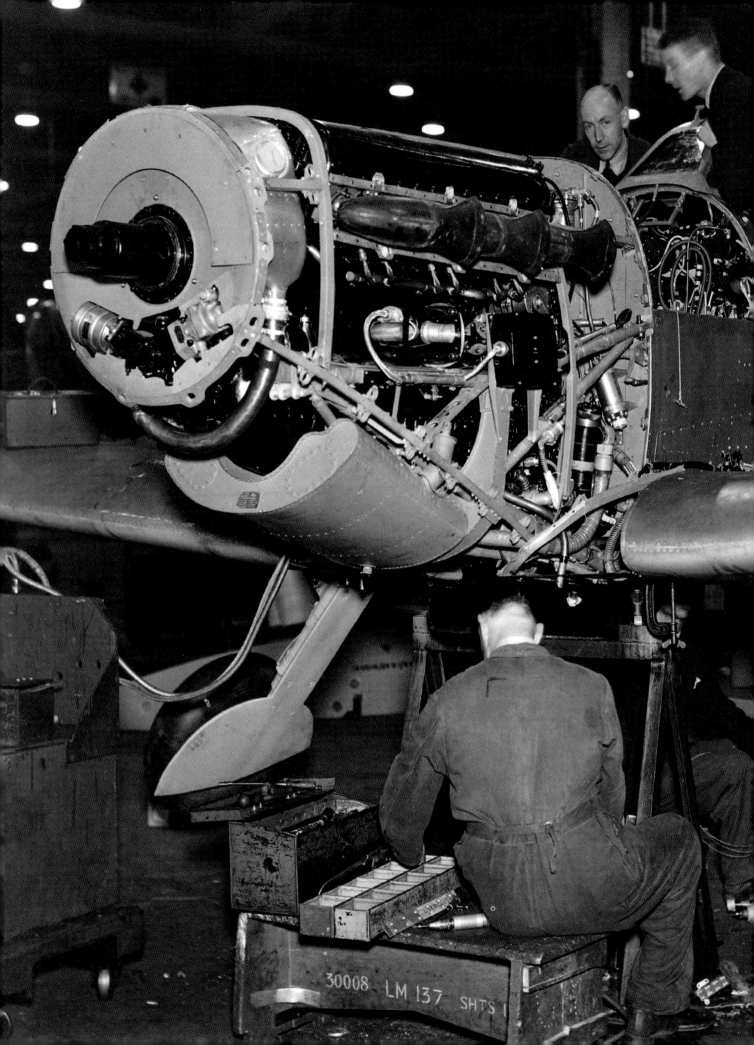

FINAL ASSEMBLY OF AN EARLY MARK SPITFIRE AT CASTLE BROMWICH NEAR BIRMINGHAM.

was then their top aerobatic pilot. He was a well-built boy, dark hair, Slavonic features, blue eyes, a firm mouth and a strong jaw. He looked a fighter and was a fighter: I would rather have had him on my side than against me. When the Germans moved into Sudetenland he watched them with the rest of his air force colleagues and then as they could do nothing with their machines under armed guard, he left his own country and asked to join the French air force. They would not accept him because France was not at war with Germany and persuaded Jicha to join the Foreign Legion. He served there under dreadful conditions until France declared war and then returned to join a French fighter squadron. He said the first time he went into action he led a flight with a Frenchman on either side in formation, they saw a squadron of Me 109s and then as he went to engage he saw the two Frenchmen turn and leave him to face the Germans alone. With the fall of France he escaped and was posted to a Czech fighter squadron in England. It was a long time before I knew and understood him, but before he left me I think if I had said to him I was going to fly into Hell itself he would have followed me."

Venda Jicha performed his job for Henshaw superbly until one day early in 1944. Alex had just returned from a test flight when he was told of a signal received that day from Air Marshal Sir Wilfrid Freeman which said, in effect, "Owing to the leakage of secret information on technical matters it has been decided to withdraw all foreign pilots from top secret factories, this instruction to be carried out forthwith." "I of course had the miserable

AIR TRANSPORT AUXILIARY PILOT
FIRST OFFICER MAUREEN DUNLOP.

job of breaking the news to Snell, Ulstad and Jicha and I didn't relish it. I called them into my office one at a time. What does one say to men who have fought for your country and with whom you have worked almost night and day for years, in conditions that weld a comradeship which is well nigh unbreakable.

"I spoke to Snell first of all: 'Ossi, the Managing Director has suggested that with the war progressing as it is I should send Allied pilots back to their own units.' Snell made my task so much easier when he butted in, 'Sure, I've been thinking the same thing. When we beat the krauts to their knees I'd like to be with my own mob as much as I have appreciated being here.' Ulstad was not so easily conned; he paused and the twinkle in his eyes told me that he knew more than he let on. 'Ja,' he answered, 'I understand what you have to do. When do you want me to go, Alex?'

"Jicha was another kettle of fish. I found I couldn't lie to him. He went as white as a sheet and for a while he was speechless. Then he spat, 'I fight for our survival, I fight for the British when they are desperate; you are my friend and I trust you, but I hate the British.' I interrupted, 'Venda, hold on a minute, you are taking this too personally. I hate it as much as you do; I think it is a very shabby treatment; it leaves a dirty taste in my mouth, but let's be reasonable. We are only small cogs in a very important machine. Someone has tripped up somewhere but this is no reflection on you. I have tried to do what I can but the ruling is quite irrevocable. I have put you up for the AFC and I am assured it will be awarded without delay. You know what I think about you and, after all,

the chances are you will be back in your own country before long.' Venda was not to be smoothed down. 'The Air Force Cross,' he spat. 'I don't want your baubles. I thought I knew the British better than this.' And with that he stormed out of my office, leaving me empty and ashamed."

"Alex, Jicha has just been killed." Henshaw learned that Venda had been posted to a maintenance unit in Scotland and was being flown from Kinloss in an Anson by an RAF squadron leader; they had crashed into the mountains, the squadron leader being killed outright. Venda, Alex was told, was comparatively unhurt; but as he attempted to crawl through the deep snow to safety, he passed out through exposure in the arctic conditions and the search party found his frozen body the next morning. "A wave of intense sadness went through me as I was told. In spite of our different nationalities, the tremendous contrast in our upbringings, and the gap in many ways of our political views, I felt Venda and I understood one another without the use of words. I respected him and I made a promise to myself that when the war was over I would go to Czechoslovakia, find his mother and sister if they were alive and let them know what a magnificent job Venda had done throughout the war, and the affection and respect he had won from all those who had got to know him."

At the end of the war, Alex Henshaw's work at Castle Bromwich ended, "It is, I suppose, rare to have one weapon that stands supreme or can contribute on its own to final victory in any war. There have of course been some near examples in our own

time: the tank in 1916, the submarine in World Wars I and II and—the most extreme demonstration—the atom bomb in 1945. But I could with little effort state a case for the Spitfire winning the last world war. At the same time that it sustained and encouraged every man, woman and child in this small island of ours, it struck not only a physical blow but a strange psychological terror in the heart of the enemy—particularly in German pilots when out of the blue flashed that elliptical wing of perfect symmetry to spurt death and destruction that survivors would remember for the rest of their lives. In spite of the numerical superiority of the Hurricane and the excellence of its performance in battle, it could not have survived alone. Neither could we, had we lost the Battle of Britain. Whatever future historians may write and say, without the Spitfire we could not have survived the largest and most bitter contest for supreme power that has yet been known in the history of the world.

"My last landing was a careful, gentle touchdown, and I taxied back to the Flight Shed as I had done so many hundreds of times before. The stillness around me seemed strange and unreal. Where were my comrades of those six long years, the loyal and willing team who had always bustled around me when I came to a standstill, the highly skilled, courageous band of pilots? A few minutes later as I drove away, leaving that lone Spitfire on the vast, empty expanse of tarmac, I sent up a short prayer of thanks for being so closely associated with this classic of our time."

Between June 1940 and January 1946 11,694 Spitfires and 305 Lancaster bombers were produced at Castle Bromwich and its dispersal factories at Cosford and Desford. 33,918 Spitfire test flights and 900 Lancaster test flights were made by Alex Henshaw and his team of pilots operating from the factory in the course of 8,210 Spitfire flying hours and 344 Lancaster flying hours. In more than five years of test flying, twenty-five pilots were involved for periods of six months or longer, most of them being RAF officers who had completed various tours of active service. In their work they experienced 127 forced landings, mainly due to engine failures. Of that number, despite the frequently poor weather conditions and the critical nature of the failures, 76 of the aircraft were landed with wheels down and no further damage. Two pilots were killed in the testing programme.

Air Commodore Allen Wheeler, an engineering staff officer in charge of experimental flying, first at Boscombe Down and later at Farnborough, was himself a test pilot. He was also the first post-war owner and operator of the Mk Ia Spitfire, AR213.

"I am often asked what sort of temperament a test pilot should have; what sort of individual makes a good test pilot?

The answer to this is quite easy. Anyone who has all the normal faculties of seeing and hearing as well as the next man, and smelling well enough to know when the 'thing' is on fire, and really wants to be a test pilot, can ultimately become a good one." An over-simplification, of course, and subject to considerable challenge, but fundamentally correct.

In the dedication of his book *Never a Dull Moment at Supermarine*, Denis Le P. Webb wrote: "In Memory of . . . the Test Pilots who found out the hard way if we had all done our sums right and built the aircraft correctly."

AN RAF SPITFIRE PILOT RUSHING TO HIS MK I IN A SCRAMBLE DURING THE BATTLE OF BRITAIN.

SPITFIRE RN201 WAS BUILT AT
KEEVIL IN 1945. THE MK XIVE
SERVED BRIEFLY WITH NO. 350
(BELGIAN) SQUADRON. IT WAS
RESTORED IN 1998 BY HISTORIC
FLYING IN THE UK.

MERLIN

It was called the Derby Hum—that incessant, seemingly never-ending drone of Merlin engines being tested, night after night, at the main plant of Rolls-Royce in the English Midlands throughout World War Two. The sound has acquired a nostalgic appeal over the years. There is no engine sound quite like it. It is distinctively Merlin.

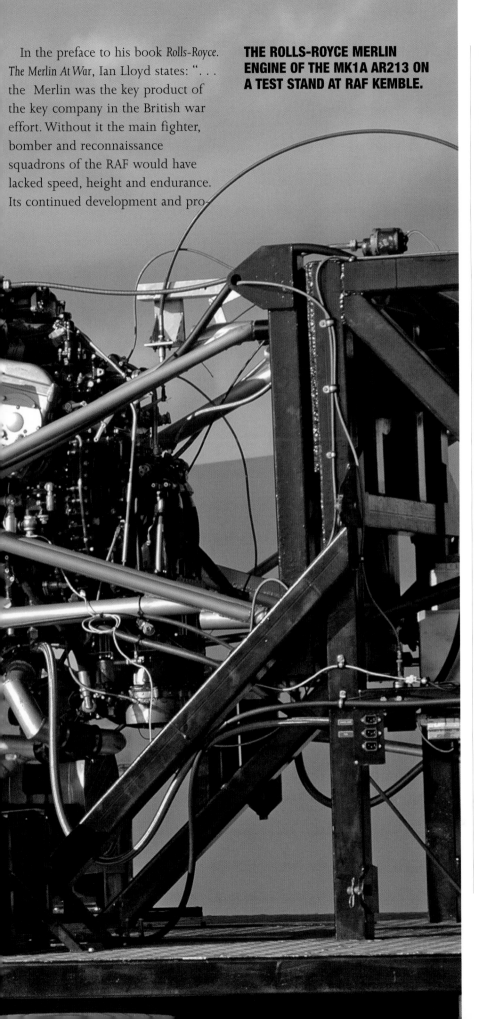

In the preface to his book *Rolls-Royce. The Merlin At War*, Ian Lloyd states: ". . . the Merlin was the key product of the key company in the British war effort. Without it the main fighter, bomber and reconnaissance squadrons of the RAF would have lacked speed, height and endurance. Its continued development and pro-

THE ROLLS-ROYCE MERLIN ENGINE OF THE MK1A AR213 ON A TEST STAND AT RAF KEMBLE.

duction was therefore a matter of the utmost importance."

Contrary to popular myth, the famous Rolls-Royce Merlin aero engine of World War Two was not named after King Arthur's wizard. Like most of the other important aero engines developed by the company, the Merlin got its name from a bird, a small falcon also known as a pigeon hawk, whose origins are, somewhat ironically, German. Falcons are raptors, or birds of prey, with thin, pointed wings enabling them to dive at extremely high speeds.

In the early 1930s, Rolls-Royce was designing the Kestrel engine which would power the RAF's Fury biplane fighter. The company hoped to develop the Kestrel into the 700 hp Peregrine, a promising evolution for the next generation of British fighter. By definition, fighter aircraft are military planes designed mainly to attack other aircraft. They are relatively small (compared with bombers), fast and highly manoeuvrable, and have evolved with state-of-the-art tracking and weapons systems for locating and destroying enemy aircraft.

Another aero engine, the Vulture, was being developed at the same time, as a 24-cylinder variant of the Kestrel in an effort to coax 1,700 horsepower from the motor. It was essentially two Peregrines bolted together with a common crankshaft. The Vulture was intended for use on bombers and other large aircraft. The company also hoped to develop a new powerplant from the basic "R" type engine used to power the Supermarine S6 racing plane of the Schneider Trophy series. All of these plans, however, left a significant gap between the 700 hp and 1,500 hp ranges, and to fill that gap R-R began work in 1933 on an entirely new, 1,100 hp aero engine they called the PV-12 (Private Venture, as no government development money was then forthcoming for it). Utilising new metallurgy and special alloys, the company committed itself to making an engine capable of generating as much horsepower as its weight and air displacement made possible at high altitude. It was a relatively small engine of just 27 litres displacement and the size was never increased through subsequent marks. The new engine first powered a Hawker Hart biplane in 1935. R-R also utilised a Fairey Battle and a Hawker Horsley as flying test beds in evaluating the engine. By the following year, the Air Ministry's requirement for a new fighter with a top speed in excess of 300 mph had resulted in the Hawker Hurricane and the Supermarine Spitfire, both of which were designed around the PV-12 engine. Prototypes of the new fighters had been successfully demonstrated and production contracts were issued in 1936. Top priority was given to PV-12 engine production and the powerplant was given the name Merlin.

In the course of the war, Merlin engines built by Rolls-Royce at Derby, Crewe and Glasgow, and by Packard in the United States, totalled 168,040 units. The various Merlin marks powered many Allied military aircraft, including the Hart, Spitfire, Hurricane, Battle, Defiant, Whitley, Fulmar, Wellington II, Lancaster, Halifax, Beaufighter II, Mosquito, Kittyhawk, Barracuda, Seafire, Mustang, and Lincoln. At the end of

AN ASHTRAY MADE FROM A MERLIN ENGINE PISTON; RIGHT: AN OPS ROOM CLOCK FROM THE FORMER RAF TANGMERE.

the war, Marshal of the Royal Air Force Lord Tedder, who had overseen development of RAF aircraft and aero engines during the Battle of Britain period, stated that the British victory in the Battle had come down to three factors: the bravery and skill of the pilots, 100-octane fuel, and the Rolls-Royce Merlin engine.

Winston Churchill believed the Merlin to be so vitally important that, in 1940, he ordered a complete set of blueprint drawings for the engine be secretly sent to the United States in case the UK were invaded by the Nazis.

The prototype Spitfire, K5054, was powered by a 990 hp Merlin C engine. The C was superceded by the Merlin II in the first production Spitfires to enter squadron service, with 19 Squadron at Duxford in

August 1938. Once the government had become convinced of the Merlin's potential, it provided subsequent funding for the project.

Arrival of the first bulk shipments of 100-octane aviation fuel from Esso refineries in the United States, Curacao and Aruba in June 1939 enabled RAF Fighter Command to convert many of its aero engines to run on the higher-performance spirit. This produced a gain in boost and upwards of 130 hp at rated altitude; this without risk of detonation in the cylinders. A further substantial power boost was provided in 1944 with the introduction of 150-octane fuel.

The Spitfire and Hurricane engines that did the work in the Battle of Britain were mainly basic single-speed supercharged Merlin Mks II and III. Merlin XIIs with higher-geared superchargers and a few Merlin XXs with two-speed super-

As good as it was (and still is), the Merlin was not perfect. The early engine was designed with an unusual "evaporative" system which provided cooling through condensers in the wings. But, with the availability of Ethylene Glycol as a coolant more efficient than water, R-R developed a smaller cooling system that eliminated the wing condensers. It was, however, vulnerable to enemy gunfire, especially when the header tank at the front of the aircraft was hit. A Spitfire or Hurricane incurring such damage would have to break off a combat engagement and land before the overheating engine seized or the flammable glycol caught fire.

Another deficiency of the early Merlin was its lack of direct fuel injection, which put it at somewhat of a disadvantage against its principal opponent in the Battle of Britain, the Messerschmitt Bf 109E with its large, fuel-injected Daimler-Benz engine. In aerial combat the Spitfire and Hurricane were often required to become briefly inverted and to dive steeply after their enemies. With its normal carburettor, the Merlin tended to cut out temporarily in these negative G circumstances, giving the Messerschmitt pilot an extra second or two to escape. The RAF pilots quickly learned to compensate for the effect by half-rolling their aircraft before pursuing the 109s in dives. In 1941, a young scientist at the Royal Aircraft Establishment, Farnborough, Beatrice "Tilly" Shilling, invented a holed diaphragm that fitted across the float chambers of the Merlin carburettor—a stop-gap modification that fixed the problem and allowed the engine to continue to function in short periods of negative G. This sufficed until the

chargers, also saw service in some Hurricane IIs during the Battle.

Continuing development of the Merlin correlated to the ever-increasing octane ratings of aviation fuel being sent to the UK from America, coupled with the increasingly efficient supercharger design work at Rolls-Royce. Early in the war Merlins were running on 87-octane aviation spirit and were providing a bit more than 1,000 hp. The next major version, the Mk XX, used 100-octane fuel to deliver 1,300 hp, with ultimate evolution bringing a combination of 150-octane and upwards of 2,000 hp by late in the war.

The early Merlins were less than reliable, but the vitally-needed engine was of such importance to the British, and ultimately the Allied war effort, that R-R initiated an urgent reliability improvement programme. Factory engineers drew Merlin engines at random from the assembly line and test-ran them at full power until they broke. As each of the test engines failed, it was dismantled and a post-mortem performed to find the specific part that had failed. That part was then redesigned to increase its strength. When the reliability improvement programme had been operating for a full two years, the Merlin had evolved into one of the most reliable aero engines in the world. Just ask the bomber aircrewmen who routinely flew their Lancasters, Halifaxes and Wellingtons at full power on seven-and eight-hour bombing missions, without any engine troubles.

By 1943, priority development of the Peregrine and Vulture engines had ceased in favour of greater emphasis on the highly successful Merlin and their programmes were cancelled.

1943 introduction of true Negative G carburettors on the Merlin. That stop-gap measure was referred to as "Miss Shilling's Orifice".

One of the most important wartime engine developments was the Merlin 60, a two-stage, two-speed inter-cooled powerplant which introduced a second centrifugal supercharger operating in tandem with the first to increase levels of compression. Able to run at up to 25 lb of boost on 150-octane fuel, the 60 engine series produced far greater high altitude performance and much more power at all altitudes. It would provide the power for the vastly improved Spitfire VII, VIII, IX, XVI, and the Mustang III, IV, and P-51B, C, D, and K.

During the Battle of Britain it became clear that many more Merlins would be needed for the fighters and bombers of the Royal Air Force than could be supplied by Rolls-Royce. The company attempted to license American production of the engine to Ford Motor Company, but Henry Ford declined to participate as he was convinced that Britain was going to lose the war. Fortunately, the Packard Motor Car Company of Detroit stepped in and signed an agreement in September 1940 to set up a Merlin production line in the U.S. The first Packard-built Merlin, designated V-1650-1, was tested in August 1941. Packard's Merlin proved an enormous improvement over its American rival, the Allison V-1710. It ultimately replaced the Allison in the Curtiss P-40 fighter and later became the powerplant for one of the best fighters of the war, the North American P-51 Mustang. The Packard Merlin was used to power Lancasters, Mosquitos and Hurricanes built in Canada and Mk XVI Spitfires built in the UK. Rolls-Royce aero engineers had not expected to be impressed by the first Packard-built Merlins arriving in the UK, but when they stripped one down they found the component tolerances were actually better than in their own hand-built and hand-finished engines. Some of the R-R engineering staff still remained sceptical about the quality of the American-produced engine which had been built by a mostly unskilled and semi-skilled female workforce. Their doubts were later dispelled when the performance and reliability of the Packard engines was experienced. Indeed, a number of Packard-introduced innovations, including easier use of interchangeable parts, were incorporated into subsequent British Merlin production. By the end of the war, 55,873 V-1650 Merlins had been built in the United States at a cost of $25,000 each.

At nearly 37 litres the Rolls-Royce Griffon engine was much larger than the Merlin. It was developed from the Type R engine the company had designed for the Schneider Trophy racing series and had been intended for low-altitude use in naval aircraft such as the Fairey Firefly. When Joe Smith took over as Chief Designer at Supermarine, following the death of R.J. Mitchell, he suggested fitting a Griffon to a Spitfire. With the resumption of developmental work on the Griffon by R-R, Smith's idea resulted in a single clipped-wing Mk IV Spitfire powered by a Griffon RG 2SM engine. The bigger, more powerful Griffon had a rough growl utterly

SPITFIRE MAINTENANCE IN THE
FIGHTER COLLECTION HANGAR AT
DUXFORD, CAMBRIDGESHIRE.

unlike the satisfying throaty roar of the Merlin, but it delivered more power. A maximum of 2,420 hp in the Griffon 101 pulled the standard production Spiteful Mk XVI to a maximum speed of 494 mph, fully loaded.

After the war, people returned to many activities put on hold for the duration of the conflict. In sports, the 1950s brought the return of championship power boat racing, and Merlin and Allison aero engines had become popular in the big hydroplane racers for their high power-to-size ratio, their availability and reliability. The splendid aero engines dominated the sport for many years until they were eventually supplanted by the even more powerful gas turbines.

Championship air racing returned with a fervour inspired by the presence of wartime fighters groomed for competition over a transcontinental course in the United States, as well as the Unlimited Class Air Racing round the "nine-mile oval" of the Reno (Nevada) Air Races. For decades the warbirds, powered by Rolls-Royce Merlin and Griffon engines, have shattered the desert calm at speeds a hundred miles per hour faster than their original specifications. R-R aero engines have claimed victory in the great majority of the Reno air races since the series began in 1964, along

AN RAF POLISH SQUADRON CANNON-ARMED SPITFIRE; BELOW RIGHT: AN RAF SPITFIRE PILOT LEAVING HIS CREW ROOM AT DISPERSAL.

with eight transcontinental distance races and twenty national championships. The victories recall the Schneider Trophy event of September 13th 1931, when a Supermarine S.6B powered by a Rolls-Royce "R" V-12 engine and flown by RAF Flight Lieutenant J.N. Boothman, won the race with an average speed of 340.08 mph to retain the Schneider Trophy for Great Britain. Two weeks later the S.6B, a floatplane, set a new world speed record of 407.05 mph.

Today Rolls-Royce is among the world's largest producers of aero engines. The company has built powerplants for airliners including the Boeing 707, De Havilland Comet (and Nimrod), the British Aerospace Trident, the Anglo-French Concorde supersonic transport, the Lockheed L-1011 Tristar, the Boeing 747, 757, 767 and 777, and the Airbus A330, A340 and A380. But the Merlin remains the most famous engine of them all.

A HURRICANE MECHANIC'S KIT USED AT RAF WESTHAMPNETT DURING THE BATTLE OF BRITAIN; RIGHT: PILOTS OF NO. 611 SQUADRON, RAF, PUSHING A NEW MK IX SPITFIRE, DECEMBER 1942.

FILM STAR

VARIOUS MARK SPITFIRES WERE GATHERED FOR USE IN THE 1968 MOVIE BATTLE OF BRITAIN WHICH WAS FILMED IN ENGLAND AND SPAIN.

The appeal and attraction of old warbird aircraft has led to their use in many motion picture and television productions, as well as product promotions since the war. Their postwar careers—in fact, their very survival and continuing existence—are directly attributable to the making in 1968 of the United Artists film *Battle of Britain*. The $20,000,000 production featured Sir Laurence Olivier, Kenneth More, Kurt Jurgens, Trevor Howard, Robert Shaw, Sir Ralph Richardson, Michael Caine, Edward Fox, Susannah York, Christopher Plummer, Ian McShane et al. The filming was done at various locations in England and Spain, mainly in the summer of 1968.

A former Polish Air Force Hurricane pilot named Ben Fisz was in the motion picture production business in London in the mid-1960s and had recently completed work on the Kirk Douglas film, *Heroes of Telemark*. In September 1966 Fisz had an idea for a large-scale movie about the Battle of Britain in which both the British and the Germans would be objectively and realistically portrayed and the British would speak English while the Germans spoke German with on-screen subtitles. There followed many months of financial turmoil and disappointment as Fisz struggled to put a deal together with the Rank Organisation to make the new film. His problems seemed nearly insurmountable when an angel appeared to him in the form of Harry Saltzman, half of the highly successful pair, Albert Broccoli and Harry Saltzman, the production wizards behind the James Bond film series. Saltzman had heard of Fisz's troubles with the Rank people and offered financing assistance which Fisz duly

THE ORIGINAL OFFICE OF AIR CHEIF MARSHAL SIR HUGH DOWDING; RIGHT: DOWDING WITH LAURENCE OLIVIER WHO PLAYED HIM IN THE FILM.

accepted. Their partnership was called Spitfire Productions Limited. Rank soon pulled out of the project and Saltzman/Fisz began discussions with Paramount. Paramount then left the negotiations, deciding there would be insufficient interest in an all-British major war picture at that time. Saltzman then went to United Artists, the group behind the Bond movies, eventually emerging with both a distribution deal and much of the financial backing for *Battle*.

Of all the films dealing with aerial combat in the Second World War, *Angels One Five*, *Reach for the Sky*, *First of the Few* (*Spitfire*, as it was titled in the United States), *Eagle Squadron*, *Fighter Squadron*, *Flying Tigers*, *Midway*, and others, only *Battle of Britain* succeeded in realistically conveying the visual experiences of fighter and bomber air crew in battle. Utilising the best cine-matic technology of the period, the film-makers captured some of the most amazing and wonderful aerial action images ever filmed and, unlike most movies of that genre, *Battle of Britain* was shot in colour for the wide screen format. History happens in colour and shooting it in colour gives it a power and immediacy that is far more difficult to achieve in black and white. Choosing to make the film in colour did, however, rule out the use of newsreel, gun-camera and other black and white footage—spliced in with that shot by the film-makers. Thus all of the images in the film had to be originally planned and shot by Spitfire Productions. It also required the use of a large assemblage of fully-functional wartime aircraft of both Royal Air Force and Luftwaffe types, along with a number of full-size, taxiable replicas. Forty minutes of magnificent and completely credi-

ble aerial combat footage also had to be photographed.

Some aspects of the air battles of France and Britain had previously been built into theatrical films such as the aforementioned *First of the Few*, but no single motion picture had ever told the entire story of the Battle of Britain. In their 1969 release United Artists achieved that aim in epic proportions. Supported by a superb musical score composed by Ron Goodwin and Sir William Walton, *Battle of Britain* takes the viewer from the Fall of France and the Dunkirk evacuation, through the Luftwaffe attacks on Channel convoys, their raids on the British radar chain and the RAF airfields in southern England, the build-up and height of the air battle and the Blitz of London, concluding with Hitler's cancellation of Operation Sea Lion, his intended invasion of Britain. It's a great story and it's all true.

During Ben Fisz's and Harry Saltzman's pre-production gyrations, the ruddy-complexioned, bowler-hatted figure of Group Captain Hamish Mahaddie, DSO, DFC, AFC and bar, a former RAF Stirling bomber pilot and a founding member of the Pathfinder Force, was scouring the world for the Spitfires, Hurricanes, Messerschmitts and Heinkels he needed for the movie. Before his involvement with *Battle of Britain*, Mahaddie had earned a reputation as the leading finder and procurer of aircraft for the movie industry. He had been responsible for sourcing the Lancasters used in the 1950s feature film *The Dambusters*.

Fisz had first approached the RAF for the British fighters the film needed, without success. But the wily Scot Mahaddie was more than up to the task. He went to work on his many Royal Air Force contacts around the globe, wherever the required aircraft

types had been operated. He had been led to believe that only about a half dozen Spitfires still existed, but when he had completed his search he had turned up more than a hundred. Hurricanes, on the other hand, were in very short supply and only a few were eventually secured for the film.

Mahaddie had also been in meetings with top people at the British Ministry of Defence and he ultimately arranged for the loan of nineteen Spitfires and three Hurricanes for use by the film company. In addition to the aircraft, the MOD provided facilities at RAF Henlow in Bedfordshire, where the necessary conversion work on the aircraft would be done, along with the assistance of twenty-one RAF fitters and tradesmen to help with their maintenance. The MOD contributed the co-operation of the Battle of Britain Flight which would participate in some of the flying sequences for the film. These arrangements amounted to an unprecedented example of co-operation by the Royal Air Force with a film organisation.

Most of the Spitfires and Hurricanes loaned by the RAF, however, had been station gate guards and many were relatively late-mark aircraft, which, if used unmodified, would not have correctly represented the Battle of Britain period. They had to be surgically and cosmetically adapted before they were ready for the cameras. Clipped wings were restored to the classic Spitfire wing shape; three-blade propellers replaced four-blade examples where possible; low-back rear fuselages were given high-back profiles with built-up wood and fabric decking, the long, tear-drop canopies were removed and the pointed rudder fins were replaced with the earlier

A STILL FROM
THE FILM BATTLE
OF BRITAIN FILM,
1968.

rounded design.

In his lengthy, painstaking hunt for the British fighters, Mahaddie pursued a number of privately-owned Spitfires. He also looked for a civilian firm capable of helping with the considerable engineering maintenance task that would keep the movie force airworthy during the shooting schedule. He settled on the Elstree-based Simpson's Aero Services, a small outfit of Merlin engine specialists near London. In searching for civilian-owned Spitfires, Mahaddie hit early paydirt when he was able to hire the use of the Mk Ia AR213, then owned by Air Commodore Allen Wheeler. The aeroplane had long been in storage at Old Warden with the Shuttleworth Collection and at Abingdon, and became one of the first of the movie Spitfires to be treated by the Simpson people, with further work being done at the RAF Henlow facility in late 1967. Other Spitfires to be rejuvenated by Simpsons included AR501, the Old Warden Shuttleworth Mk V, the Mk IX MH434, then owned by Tim Davies, and two former Irish Air Force two-seat trainers, MJ772 and TE308, owned by Tony Samuelson. Additional flyable Spitfires made ready for service in the filming were the Mk IX MK297 which had recently been acquired by the then Confederate (now Commemorative) Air Force in Texas, the Griffon-engined Rolls-Royce Mk XIV RM698, and a French-owned aeroplane, the Mk IX MH415, all of them prepared for their movie roles by Simpson's.

It was the German aeroplanes that presented Mahaddie's greatest challenge. Where was he to find enough He-111 bombers and Me-109 fighters to field the mass formations called for in the shooting script? The answer came in a conversation Hamish had with former Luftwaffe General Adolf Galland, on whom the character Falck is based in the film. Galland suggested that Mahaddie contact the Spanish Air Force which, in the 1960s, still operated Spanish-built versions of the He-111 bomber designated CASA 2.111, as well as the Hispano HA 1112 Buchon version of the Messerschmitt 109 fighter.

Enlisting the help of RAF Group Captain R.L.S. Coulson, the British Air Attaché in Madrid, Mahaddie learned that the entire force of Spanish Messerschmitt fighters was being dismantled at Tablada Air Base near Seville, except for seven aircraft that were in good shape and were shortly to be sold to the highest bidder in a sale called a *sabasta*, a sealed-bid auction. After a good deal of anguish the Scotsman was able to acquire the Messerschmitts as a job lot for a little over $2,000 each. As for the Spanish Heinkel bombers, Mahaddie mustered all of his considerable charm and persuasive skill during a lunch with the Spanish Minister for Air, and was frankly surprised at the Minister's response days later to a follow-up letter from Group Captain Coulson. The Spaniard replied that his government had not only given approval for the film-makers to use the bombers in the air and on the ground, but all expenses including fuel and maintenance would be born by the Spanish government, apart from the painting and marking of the aircraft for the film. The gesture saved the film production company more than $400,000 and made more than fifty Heinkel bombers available to them for their "Luftwaffe".

When his work was done Hamish Mahaddie had provided far more aircraft than the number actually required for the *Battle of Britain* filming. Many of the flying and non-flying planes served as set decoration to fill out scenes as needed. The flying aircraft used in the shooting schedule were: nine Spitfires, three Hurricanes, eighteen "Me 109s", three "Heinkel He-111s", two Junkers Ju 52 (CASA 352) transports and, as photography platforms, one helicopter and one North American B-25 Mitchell bomber. Additionally, the Battle of Britain Flight had made three Spitfires and one Hurricane available for the picture.

On March 13th 1968, shooting of the first of nearly 5,000 separate shots began at three in the afternoon on Tablada airfield. It was part of the opening title sequence of the film when German Feldmarschall Erhard Milch inspects two rows of Heinkel bombers.

Money problems were not the only ones to plague the production company in that difficult spring and summer of 1968. The weather in the main shooting locations, in both Spain and England, proved wholly unsympathetic to the moviemakers. In Spain especially, rain and deteriorating weather conditions crippled the shooting, leaving the production perilously behind schedule. The entire production was in jeopardy. Then, roughly halfway through the filming in Spain, Spitfire Productions was informed by the Spanish Air Ministry that all the borrowed bombers of the "Heinkel" air fleet were suddenly required for a ceremonial flypast during a forthcoming NATO exercise in the Atlantic. This was

an additional disaster for the shooting schedule, with the prospect of even greater delays while the green and black camouflaged bombers were repainted in their former Spanish Air Force scheme. The estimated cost of stripping and repainting the planes was to be billed to the film-makers at about £1,000 per aircraft. Guy Hamilton, the film's director, flatly refused to accept this latest blow to the production, protesting vehemently to the Spanish authorities who, in the end, agreed to compromise and allow the flypast to proceed with the bombers retaining their movie Luftwaffe paint. Hamilton was forced to accept the brief added delay to his shooting schedule. In early May, when filming was finally completed in Spain, it took fully five days to ferry all of the Messerschmitts, two Heinkels and the B-25 camera ship to Duxford, near Cambridge,

where shooting would continue throughout the summer and well into the autum.

The B-25 Mitchell bomber used as the principal camera platform aeroplane for the production was flown by John "Jeff" Hawke, a former RAF pilot whose previous cinematic activity had included piloting one of the De Havilland Mosquito bombers in another United Artists air war film, 1964's *633 Squadron*, which starred Cliff Robertson. Hawke and fellow pilot Duane Egli racked up 300 flying hours in the Mitchell to put the essential forty minutes of superb air fighting and formation sequences on film for *Battle of Britain*.

The primary shooting locations in England were Duxford, nearby North Weald, and Hawkinge, on the Kent coast. All of these sites had been actual Battle of Britain airfields and, in

1968, still at least partially resembled the way they had looked in the summer of 1940. Debden, an airfield slightly southeast of Duxford, had been an RAF station during the Battle and, like Duxford, had been used later in the war by the 4th fighter group of the U.S. Eighth Air Force. It served as an additional operating base for the film production company.

Duxford was the most adaptable of these fields, providing locations for the French airfield shown in the early Battle of France scene; for Duxford airfield itself in the Battle; and for the South Downs Flying Club sequence. The Duxford airfield and base had been deserted by the RAF in 1961 and remained unused, in the care of the Home Office, until the mid-1970s when London's Imperial War Museum began using the site for exhibit storage. In time, it became home to much of

LEFT: ROBERT SHAW IN THE BATTLE OF BRITAIN FILM; BELOW: ROBERT STANFORD TUCK AND ADOLF GALLAND ON A SET OF THE FILM,

the museum's large static aircraft collection and has since hosted a number of private aircraft collections including Ray and Mark Hanna's Old Flying Machine Company, and Stephen Grey's The Fighter Collection. It is now the venue for several annual warbird-focused airshows, with the July Flying Legends ranking among the world's best air displays.

Since *Battle of Britain* was made, these historic old airfields have changed somewhat. Duxford, unlike the others, retains most of its hangars and its station buildings, and has added some major new buildings which, while offering marvellous exhibits, are nevertheless certainly not in character with the original look of the base. North Weald, only minutes down the M11 motorway towards London, has lost its pre-war hangars and most of its other structures, but has kept some

of its Second World War T-2 type hangars and still hosts some of Britain's other important warbird operations. And over in Kent the old Hawkinge aerodrome, the only all-grass airfield of the lot, has given way to a new housing estate.

In the course of filming in England a number of distinguished visitors and members of the press arrived at the Duxford site. On May 26th Lord Dowding, who had led RAF Fighter Command in the Battle of Britain, came to the field. At that point he was confined to a wheelchair. Veterans of the Battle present with him were Al Deere, James "Ginger" Lacey, Douglas Bader, Peter Townsend, Johnnie Kent, Tom Gleave and Robert Stanford-Tuck. Before visiting the airfield Lord Dowding had called at Pinewood Studios near London where he met Sir Laurence Olivier who played

Dowding in the movie.

Hamish Mahaddie served as Chief Technical Advisor for the movie and, being an old bomber man, thought it essential that one or more RAF Battle of Britain fighter aces should be on the technical advisory team. He brought in Ginger Lacey and Bob Stanford-Tuck to fill that role. To represent the German Air Force in the film, the company appointed General Adolf Galland, who had been Luftwaffe General of the Fighter Arm and one of the highest-scoring and most accomplished German fighter aces of the war.

Cinematography for *Battle of Britain* had been completed in the autumn of 1968. When this author first visited Duxford airfield in the summer of 1969, it was deserted. As I wandered through the wartime air raid shelters at the edge of the southern hillside bordering the field, I found a discarded

PETER TOWNSEND IN A
SPITFIRE DURING THE FILMING OF
BATTLE OF BRITAIN.

page from a copy of the film's shooting script, a tattered relic of the "second Battle of Britain" fought in those same skies twenty-eight years after the first one. I called on a farmer whose land adjoined the airfield and he told me that he had witnessed both "Battles" over his farm and that, quite frankly, he thought the movie version had been more realistic than the actual Battle.

By the end of the filming, the various aircraft used in the production had logged a staggering total of more than 5,000 flying hours and they had formed the world's "35th largest air force". AR213 had played a significant part in the movie, appearing in several scenes and wearing at least seventeen different squadron codes. She was well cared for and, when returned to her then-owner, Air

Commodore Allen Wheeler, he was pleased that his aeroplane had not only been rejuvenated but had been fitted with a new Merlin engine by Simpson's at Henlow and was, in his words, "in superb condition".

The movie premiered in London on Battle of Britain Day, September 15th 1969, at the Dominion Theatre, Tottenham Court Road and thereafter around the world. The reviews were mixed. The opinion of many was that the film was overlong, did not come up to expectations, but that it *did* catch the atmosphere of those summer days in 1940 and *did* deliver gripping dogfights accompanied by stirring music. Despite the criticisms, however, there seems to be general agreement that *Battle of Britain* is among the best-made, most authentic and visually pleasing war films ever.

Piece of Cake was the most important television production in which the Spitfire AR213 has appeared. It was based on the book of the same title by Derek Robinson, described on its jacket as "a moving and engrossing novel of World War II by the author of the acclaimed *Goshawk Squadron*—the powerful, quietly savage, funny and heart-piercing story of a group of young RAF pilots, mere boys of 19 and 20, during the year from September 1939 (the beginning of the so-called Phony War, when fighting men were filled with chivalric fantasies of honour and derring-do, and life in uniform was a luxurious escapade), to September 1940: the deadly realities of the Battle of Britain and the huge Luftwaffe raid on London that was the culmination of Hitler's airborne Blitzkrieg."

The television series proved controversial, particularly among veteran Royal Air Force airmen who had flown in the Battle of France and/or the Battle of Britain. Many of them were incensed at the portrayals of the Hornet Squadron pilots in the series, feeling that they, the vets, had been made to look like a lot of thoroughly unpleasant types. In a 1989 issue of *Aeroplane Monthly* magazine, Battle of Britain ace Roland Beamont wrote: "There was no sense of defeatism at any time in any of the squadrons that I saw in action, and a total absence of the loutishness portrayed in *Piece of Cake*. It would not have been tolerated for a moment . . . The prevailing atmosphere was more akin to that in a good rugby club, though with more discipline. Nor was there any sense of 'death or glory'. RAF training had insisted that we were there to defend this country, and now we were required to do it—no more and no less. There was no discussion of 'bravery' or 'cowardice'. People either had guts or they did not—but mostly they did. But we knew fear, recognised it in ourselves and in each other, did our damnedest to control it, and then got on with the job."

In a piece about writing the novel, Robinson said of his fighter pilot characters, "They became a very individual group, all of them special if not all admirable; and contrary to the myth created by the cinema, the best fighter pilot was not necessarily the most attractive person; sometimes quite the reverse." Robinson pointed out that "All war requires killing; what made the fighter pilot different was that he killed in a spectacular way, and he did it in a uniquely beautiful setting. In 1939 he was one of a few special young men suddenly given hitherto superhuman powers:

he could break free of the Earth, climb higher than the Alps, turn, roll, dive, all at astonishing speed; and, above all, by a few seconds pressure of the thumb he could wipe out another aeroplane just as great and complex as his own; could make it blow itself to pieces and vanish in a brief, bright flowering of flame."

The war correspondent Alan Moorehead wrote of RAF pilots in his 1944 *African Trilogy*, "Most of them were completely unanalytical. They were restless and nervous when they were grounded for a day . . . They lived sharp vivid lives. Their response to almost everything—women, flying, drinking, working—was immediate, positive and direct . . . They made friends easily. And never again after the speed and excitement of this war would they lead the lives they were once designed to lead."

Clearly, there were no Spitfires involved in the Battle of France, only Hurricanes. Derek Robinson used Hurricanes in the novel and was, in that and most other respects, historically correct. The makers of *Piece of Cake* certainly wanted to use Hurricanes in their production, but there were only three airworthy Hurricanes in the entire world in early 1988 when they would be filming, and of those one was in a Canadian museum. The other two were the property of the RAF, which strictly limits the number of hours it allows its historic aircraft to be flown. Being determined to show several British fighters in the air for the series, the producers elected to use Spitfires and hired them from private owners. Principal among these aircraft was Mk Ia AR213, then owned by the Hon. Patrick Lindsay, International Director of Christie's, the auction

house. Lindsay had bought AR213 from Air Commodore Allen Wheeler and continued to fly it from Wycombe Air Park where it was maintained by Personal Plane Services. As mentioned earlier, this aircraft had also featured in the *Battle of Britain* movie and in other film work. Then the world's only flying Mk I, it was unique among the six Spitfires employed in *Piece of Cake* as an example of the variant used in that early war period, with the exception of having a four-blade propeller instead of its original three-blade prop.

The other Spitfires utilised in *Piece of Cake* were MH434, a Castle Bromwich-built Mk IXB that had served with No 222 (Natal) Squadron at Hornchurch in 1943 and been credited with downing two Fw 190s and the shared kill of a Me 109F; ML417, a Mk IXE that had served with No 443 (Canadian) Squadron at Ford in 1944

and was credited with shooting down two Me 109s with two Fw 190s as "probables"; NH238, another Castle Bromwich-built aeroplane, which, in the 1970s, had been operated by the Confederate Air Force (later changed to the Commemorative Air Force) in Texas and been returned to Britain in 1984 when it was purchased by Doug Arnold; PL983, a Mk XI photo reconnaissance Spitfire that had been flown from Blackbushe in 1944 by the 2nd Tactical Air Force; and MK297, another Mk IX that had been a part of the Confederate Air Force and had appeared in the *Battle of Britain* film.

Ray Hanna, a former leader of the RAF Red Arrows display team and owner of the Old Flying Machine Company at Duxford, was appointed chief pilot for *Piece of Cake*. His son, Mark, also a former RAF fast jet pilot, was deputy chief pilot on the

production. Stephen Grey, owner of The Fighter Collection at Duxford, and airline pilot Michael "Hoof" Proudfoot, also flew the Spitfires, as did Peter Jarvis, Carl Schofield, Brian Smith, John Watts and Howard Pardue. Proudfoot flew AR213 in most of the *Piece of Cake* filming. The required German aircraft for the series were secured from the CAF in Texas, which provided their CASA Heinkel He-111 bomber and two Spanish-built Buchon "Me-109"s, also veterans of the *Battle of Britain* film. They were flown by Nick Grace, Reg Hallam and Walther Eichorn. The B-25 was flown by Vernon Thorp, Anita Mays, Jack Skipper and Walther Wootton; the Harvard by John Romain; the Junkers by Peter Hoare; and the helicopter by Michael Malric Smith. Finding appropriate shooting locations for *Piece of Cake* was the task of location manager Mike Hook, whose primary assignment was a site for the French château and airfield of the early episodes. The series had to be shot in England and the trick was to find a period country house that would pass for a château. He recalled visiting Charlton Park, a Jacobean mansion near Malmesbury in Wiltshire, years earlier and thought it might meet the requirement. Owned by the Earl of Suffolk, Charlton Park also happened to come with a lovely, if too-short, grass runway which could be lengthened to accommodate the Spitfires. Lord Suffolk and his family were living in quarters converted from a former stud farm on the property and the mansion itself had been turned into apartments. Most of the residents proved quite obliging when asked to put up with the disruption of a visiting film crew, the occasional explosion and cloud of smoke, and half a dozen Spitfires flying from their lawns. For many of them it was a treat.

With Ray Hanna at the controls, the first of the Spitfires arrived in November 1987, setting down on the heavily rain-soaked grass strip. When the filming was to begin the following May, however, the landing field was completely dry, the grass beautifully filled in and firm. The other prime filming location was far less satisfactory.

The production company needed a location site on the downs above the cliffs somewhere along the south coast of England and they chose Gayle's Farm at Friston, the location of a former wartime RAF/airfield first used in 1936. During the war it had served as an emergency landing field for damaged Allied aircraft returning

from raids over the Continent and, briefly, as a base for Spitfire and Hurricane squadrons. Never graced with any permanent facilities, the Friston site was suitably dressed with small army tents and wooden huts as it had been in the war. *Piece of Cake* production manager Jake Wright found the location, an overgrown pea field, in 1988, and the lengthy job of clearing the crop and the many thousands of flint stones covering the area got under way. The stones represented a genuine hazard to both film crew and aircraft and it took a long and concerted effort on the part of thirty stonepickers to make the runway safe and useable.

The utterly barren site was suitably "furnished" by the series art director, Jane Coleman, with a Nissen hut, dozens of tents, three wooden "spider huts", and twelve old air force vehi-

cles including a fuel bowser and an ambulance. In addition, the company had six replica Spitfires constructed to fill out the required complement of squadron aircraft for the various ground scenes.

For the final airfield location, representing the fictional "Kingsmere" (an RAF fighter station in Essex), Wright chose South Cerney, a pristine pre-World War Two RAF station with an all-grass airfield, all four original hangars and most of its old buildings still standing, including the thirties flying control tower. In addition to the sites mentioned, the company was able to use the airfields at Duxford, Headcorn in Kent, and Little Rissington in Gloucestershire, as bases for the aircraft.

Abandoned by the air force in 1969, South Cerney, which is just to the south of Cirencester in Gloucestershire, had become home to an Army Royal Corps of Transport regiment and was used for gliding at weekends. The place looked right and was a perfect setting for the Spitfires. *Piece of Cake* director Ian Toynton: "Even parked on the grass, a Spitfire has considerable power and presence. They are so charismatic, they dominate any scene they are in. I did not want the Spitfires to upset the balance of the drama. So when we came to editing *Piece of Cake*, I tried to ensure that the audience would be left feeling they wanted to see more of the aircraft."

Except for a few regrettable historical inaccuracies (like the use of Spitfires instead of Hurricanes in the Battle of France), most reviewers and watchers rated *Piece of Cake* rather highly, some comparing it favourably with Derek Robinson's novel, which many people consider one of the best air war books ever.

YOU NOT ONLY GET A CAR AND A GIRL BUT A PIECE OF HISTORY. That was the headline of a 1970s magazine print ad for the Triumph Spitfire 1500 sports car. It was also the tag line for the related television commercial. Both promotions featured AR213 and the magazine ad carried a small insert photo of James Harry Lacey with the caption: GINGER LACEY, BATTLE OF BRITAIN ACE, (18 KILLS) GROWLS SPITFIRE. In the tv commercial, both Spitfires are fired up and then pace each other across the grass airfield at Wycombe Air Park, formerly the wartime RAF Booker aerodrome. It was a wonderful example of the use of nostalgia in an advertising context and apparently succeeded in expanding sales for the little Triumph car.

The copywriter's product for the print ad read: "Way back then a new car flashed on the racing world. Spitfire! Honouring the plane that saved Britain. It has done the name proud, racking up three national Class F championships, driving British Leyland to more national production victories in Sports Car Club of America competition than any other manufacturer. For thousands this lovely two seater was their first sports car. It took an uncanny grip on owners. Says one buff: 'I now have a Ferrari. But I still think back to that damn Spitfire.' Straight line integrity, then till now. Still the same throaty sound, the tight circle, the snug seat, the rollicking ride, the intimacy, man and machine—pure Spitfire. This year, a bigger 1500cc engine, 2 inch wider track, higher 3.89:1 axle ratio, larger 7 1/4 inch clutch. We look for new tracks to conquer. There it is. A car, a girl, a piece of history. Like your first love, you'll never forget your first Spitfire."

LEFT: SPANISH-BUILT 'HEINKEL' BOMBERS IN A STILL FROM THE BATTLE OF BRITAIN FILM; BELOW: A CLOSE-UP OF ONE THEM AT THE DUXFORD SHOOTING LOCATION; FAR LEFT: THE COCKPIT OF SPITFIRE AR213 PRIOR TO RESTORATION.

PM631, A PR XIX BUILT BY VICKERS AT READING IN 1945. IT HAS BEEN MAINTAINED AND FLOWN BY THE BATTLE OF BRITAIN MEMORIAL FLIGHT FOR MORE THAN HALF A CENTURY.

WARBIRD

ABOVE: AN IMPRESSIVE ROW OF SPITFIRES AND A FEW HURRICANES, AT THE ANNUAL FLYING LEGENDS AIR SHOW, DUXFORD; BELOW: P-51 MUSTANGS AT THE SAME SHOW.

A warbird may be defined as an aircraft that was originally designed for military use and is now in private ownership. There are exceptions. The Battle of Britain Memorial Flight aircraft, for example, are owned and operated by the Royal Air Force. Most warbirds, however, have been acquired since the end of the Second World War by private collectors and enthusiasts who are interested in preserving these rare machines.

For many years the worldwide warbird population grew slowly. Non-airworthy aircraft were gradually restored and made flyable and wrecks were occasionally discovered and acquired for restoration. Two events occurred that dramatically changed the pace and scope of the movement: production of the *Battle of Britain* movie in the late Sixties led to the purchase by private collectors of WW2 aircraft used in the film, and the collapse of the Soviet Union in 1991 opened the way for the discovery and acquisition of many more warbirds by westerners. Both these events brought greatly increased interest and enthusiasm among warbird aficionados, along with escalating prices for the aircraft, and today the market is hotter than ever.

Among the pioneers of the movement was Albert Paul Mantz, one of the leading movie stunt pilots from the 1930s through the 1960s. His amazing career included many spectacular scenes in which his flying and/or aerial direction added immeasurably to such films as *Hell's Angels*, *I Wanted Wings*, *Twelve O'Clock High*, *The Best Years of Our Lives*, *For Whom The Bell Tolls*, and *Flight of the Phoenix*. For *Twelve O'Clock High* he staged and flew the memorable belly-landing of a B-17 bomber—footage that has been re-used in a

number of other productions since the 1940s' hit film.

In the spring of 1946, when most people wanted to put the Second World War out of their minds and get on with their lives, Mantz sought and found an opportunity to add significantly to his existing stable of old aeroplanes—some for use in the revived air racing championship contests, the rest for motion picture work. He bought a small air force of nearly 500 fighters and bombers from the American Reconstruction Finance Corporation for $55,000— aircraft which had cost the American tax payers $117,000,000. Mantz knew that the scrap value of the aircraft aluminium alone was worth three times the purchase price, and that the aviation gasoline still in the fuel tanks was worth more than he had paid for the entire lot.

The warplanes covered the runways, aprons and revetments of the old Stillwater, Oklahoma airfield. Mantz now owned seventy-five B-17 Flying Fortresses, 228 B-24 Liberators, ten B-25 Mitchells, twenty-two B-26 Marauders, eight P-51 Mustangs, six P-39 Airacobras, ninety P-40 Warhawks and thirty-one P-47 Thunderbolts, along with several other types. His only problem was getting all of the aircraft out of the Stillwater facility by the government-imposed deadline.

The task seemed impossible. He appealed to an old friend, an executive at Warner Brothers, the Hollywood studio that up to then had made the largest number of aeroplane movies, suggesting that Warners could save a lot of money by having its own aviation department. He offered the moviemakers several of his aircraft for use in their films at half his normal

rate if they would pay the freight to move the planes to the west coast. Warners declined the offer, preferring to remain strictly in the movie business and not in the junk business. Ultimately Mantz had to accept that he could not bring all his newly-acquired aeroplanes out to California and was forced to sell most of them for scrap. He did, however, manage to move twelve of the best examples to his southern California base. For a brief moment he had been in command of an air force whose aircraft inventory was exceeded only by those of the U.S., Britain, the Soviet Union, France, Australia and Canada.

Mantz restored and operated his small fleet of military aircraft until July 1965. In the early morning of the eighth he was scheduled to fly an oddly-contrived plane constructed from "the wreck" of a Fairchild C-82 Packet, also known as the Flying Boxcar, for the film *The Flight of the Phoenix*. In the heat of Arizona's Buttercup Valley, Mantz was to double for Jimmy Stewart as pilot of the strange single-engined craft as it struggled to fly in the film's climactic scene. As he banked and descended into the valley, the aircraft struck a small hillock. Mantz applied full throttle trying to recover, but nosed in and flipped end over end. The plane broke apart and when the wreckage of the cockpit section came to rest, Mantz slid out, crushed. It was the end of one of the great movie flyers and a pioneer in the warbird phenomenon.

They were built as fighters, bombers, trainers, cargo transports, observation planes, flying boats, etc. They performed all the functions required of them in wartime and, like many of the men and women who built them, the war planes were quickly discarded when peace came. Few people who had endured and survived the war years wanted to be reminded of those terrible times. But gradually, interest was aroused in these old aircraft among a handful of pilots and aviation enthusiasts. Most felt that at least a few examples of the more important types should be saved, not merely as static museum displays, but as flying machines to be celebrated for the enjoyment and education of the general public and future generations. They were not out to glorify war, though they were sometimes criticised for doing so.

It was not a rapidly growing movement. It happened slowly, over many years. A distinct warbird community developed, encompassing men (and some women) whose love of the lines, the sight, sound and smell of the old planes united them in a common bond to restore, preserve and fly them for their own gratification and in honour of the people who had built, maintained and flown them in wartime. Thanks in large part to their contribution to the war effort and their sacrifice, the Allies achieved the inevitable victory that Churchill and Roosevelt had spoken of during the war. For many people today, the sight and sound of the exciting; powerful old warbirds at air shows around the world serves as a perfect tribute to those associated with them in the war years.

The warbird community is now global in scope, but some of the most concentrated restoration and display activity takes place in England at Duxford Airfield, the Imperial War Museum facility near Cambridge, and in the United States at Chino, southeast of Los Angeles. Of considerable

interest and impact at Duxford is The Fighter Collection of Stephen Grey, among the finest collections of mainly single-engine fighter aircraft of the Second World War. The force behind what is perhaps the world's best air show, *Flying Legends*, TFC restores, maintains and operates a range of impressive aircraft in the largest collection of airworthy warbirds in Europe, and has flown some of them in major motion pictures including *Memphis Belle*, *Pearl Harbor*, and *Dark Blue World*. Featuring prominently in the TFC line-up is the largest group of Grumman "cat" types in Europe: an F4F Wildcat, an F6F Hellcat, an F7F Tigercat and an F8F Bearcat. Of particular distinction at TFC is the North American P-51D Mustang, *Twilight Tear*, which served in WW2 with the Duxford-based 78th Fighter Group late in the war and shot down three German aircraft, two of them Me 262 jet fighters. *Twilight Tear* was the mount of Lt Hubert Davis of the 83rd Fighter Squadron, 78th FG. Equally noteworthy is the TFC Hellcat. It was flown in the Pacific campaign by U.S. Navy Lt Alex Vraciu while serving with VF-6. In the "Great Mariana's Turkey Shoot" Vraciu destroyed six Japanese dive-bombers in eight minutes, and achieved a total of nine aerial victories in the Hellcat. He ended the war with nineteen victories.

Another aircraft of prominence at Duxford is ML407, the T Mk IX Spitfire owned and flown by Carolyn Grace, the first woman to solo on a Spitfire since the 1940s, the first woman to own and fly a Spitfire, and the only qualified female Spitfire pilot in the world. ML407 was built as a Mk IXc at Castle Bromwich in 1944 and began wartime service with 485 (New Zealand) Squadron, delivered

ABOVE: A P-47 THUNDERBOLT OF THE DUXFORD-BASED FIGHTER COLLECTION IN THE 1990S.

THE B-25 'GRUMPY' WAS ALSO PART OF THE FIGHTER COLLECTION FOR A TIME.

ABOVE: A DOUGLAS AD
SKYRAIDER AT DUXFORD;
BELOW, THE COCKPIT PANEL OF
A DE HAVILLAND MOSQUITO.

from the factory by another high-flying female, Air Transport Auxiliary pilot Jackie Moggeridge. It flew 137 operational sorties with the squadron and served with other RAF squadrons (Free French, Belgian, Polish and Norwegian) as well.

By the end of hostilities ML407 had flown more than 200 combat hours. After the war the Spitfire was converted by the manufacturer to a two-seat T Mk IX configuration and, like so many other surviving Spitfires of the time, was drafted to fly in the United Artists film *Battle of Britain*, in 1968. Carolyn's husband Nick, an engineer, acquired the aeroplane in 1979 and spent five years restoring it before he and Carolyn took it up for the first time in April 1985. Tragically, Nick was killed in a car accident in October 1988. Of the Spitfire, he had once said: "I have put a piece of this country's heritage back in operation. It's been around since April 1944, it's survived 174 operational sorties, apparently with hardly any damage of any description. It's had a fairly graced life if you like, and I hope that it will keep it up. Certainly it will outlast me—if luck is on its side—and another few generations after that." It had been Nick's ambition for Carolyn to fly ML407 solo and she decided to fulfill that aim for him and for herself. On July 17th 1990, after only four hours of dual instruction in the aircraft, she did that solo flight, appropriately at White Waltham airfield, west of London, where many female ATA pilots had flown Spitfires during the war. Today Carolyn flies ML407 at both public and private air displays in the original paint scheme and markings of 485 Squadron.

From a combat report of Flying

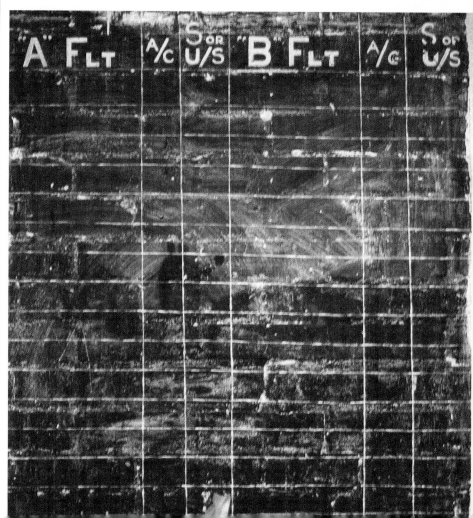

Officer J. A. Houlton, one of the RAF pilots who flew ML407 during the war: "I was leading Blue section, patrolling at 5000 ft 15 miles W. of Caen and flying E. E/A [enemy aircraft] were reported by 222 Squadron above the canal S. of Cabourg and I climbed to the S.E. and intercepted 12+ Me 109s flying S. below cloud. They were being attacked on the port side by 5 or 6 Spitfires. I singled out the e/a flying on the starboard of the formation, and closing to 300 yards, opened fire from line astern. I saw heavy strikes, pieces flew off and black smoke came from his engine. He porpoised in and out of cloud a few times and as he came out I got in one or two more bursts without effect; on the last occasion, however I followed him up and hit his slip stream in the cloud. I fired again and the black smoke came past me in the cloud. We broke cloud simultaneously and I saw him go down in a shallow dive to crash in a wood. Just before crashing an object came away but no parachute. I claim 1 Me 109 destroyed."

A much larger resident at Duxford is the Lockheed/Vega-built B-17G *Sally B*. The four-engine heavy bomber of WW2 is among the last of the 12,731 B-17s produced and was delivered from the Burbank, California factory too late to see wartime service. For several years after the war the aeroplane was part of the French Institut Géographique National, and in 1975 was sold, brought to the UK, renamed *Sally B* and launched on the air show circuit where she has been performing almost continuously ever since. In 1989 she appeared in the film *Memphis Belle*, in the title role. She is another restored warbird owned by a woman, Elly Sallingboe.

Warbird restoration and operation in the UK is by no means confined to the historic airfield at Duxford. At Coningsby, over in Lincolnshire, the Battle of Britain Memorial Flight of the Royal Air Force restores and operates Spitfires, Hurricanes, a Dakota transport, two De Havilland Chipmunks and one of the only two remaining airworthy Lancaster bombers in the world. The Memorial Flight came into being at RAF Biggin Hill, Kent, in 1957 as the Historic Aircraft Flight. The aircraft of the Memorial Flight are in action each air show season in a gruelling schedule of about 600 individual aircraft sorties. A commander of the Flight, Squadron Leader Paul Day, commented as he sat in the cockpit of a BBMF Spitfire: "The undercarriage lever . . . where that is positioned is really undesirable. On the pilot's right, it means a change of hand from the control column to operate. Not a good idea." Slapping the curved side pieces of the front Perspex, he points out that they too, are cause for concern: "The glass is contoured and not of good optical quality. That plays all sorts of tricks from a cockpit where visibility should be of paramount importance, but turns out to be a real problem. For instance, that whacking great engine obscures most of the sky. At five hundred feet, for example, you are blind for three miles ahead of you. Then, what is going on beneath those wings? Miles of airspace obscured by them. Looking behind . . . well, that's a real labour of love. There is a teeny make-up mirror for rearward vision, but that's of little practical value. As for the instrument panel . . . not badly laid out, but all the engine instruments are grouped to the right. They aren't

terribly eye-catching. In a tight spot all sorts of things could be going on there which you might not notice. The system for arming and firing, though, is simple, effective and good." Summing up, and despite his dislikes, Day felt that "Mitchell had got it right".

Considering the technology of the 1930s, it is a fair view, and it compares favourably with the Squadron Leader's later evaluation of the Messerschmitt Me 109. Day spoke from the cramped cockpit of an Me 109 after a trial closing of the heavily-framed hood: "Well, good grief! It's terrible. Not for the squeamishly claustrophobic. There is twenty-five per cent less working room in here and up front the vision is even worse. All you can see is Krupp of Essen. Looking behind is a hundred per cent worse than in a Spit." Squadron Leader Day has never flown the Messerschmitt, but in his verdict he is absolutely certain: "I wouldn't choose to go to war in one given that the opposition had Spitfires."

In 1944, four years after the death of thirty-one-year-old RAF pilot Richard Ormonde Shuttleworth in a flying accident, his mother founded a permanent memorial to him. She placed the Old Warden Estate in a charitable trust for the purpose of agricultural and aviation education, her son's primary interests. The Shuttleworth Collection opened to the public in 1963 and its regular air displays have attracted hordes of visitors from around the world.

Shuttleworth had founded the collection at Old Warden in 1928, constructing workshops and a hangar beside a small and lovely grass airfield. A racing driver as well as an aviation enthusiast, he began the aircraft collection with a De Havilland Moth

AN F-86 SABRE JET FIGHTER AT THE
FAIRFORD, GLOUCESTERSHIRE AIR SHOW.

which he used to fly between Old Warden and the Brooklands racing circuit near Weybridge. With the outbreak of war in 1939, the Shuttleworth aircraft and vehicles were stored, and private activity at the airfield ceased for the duration. The RAF used the facility for the assembly and flight testing of training aircraft throughout the war years. Since the war, more hangars and workshops have been added and the aircraft inventory has grown to around forty machines ranging from

a 1909 Bleriot to a Mk Vc Spitfire. Old Warden is renowned for providing many displays throughout the summer season in which the aircraft perform relatively near the crowd line, and for being an especially pleasant venue for the spectators.

Personal Plane Services at Wycombe Air Park, Booker Airfield near Marlow in Buckinghamshire, has been in the business of magnificently restoring aircraft, including many warbird examples, since the late Doug Bianchi founded the firm in 1967. Now owned and

operated by his son—sports, aerobatic and warbird pilot Tony Bianchi—PPS is widely regarded as one of the best operations of its kind in the world. A typically impressive example of the company's efforts is the Mk Ia Spitfire AR213 which was beautifully restored recently by the PPS team. The thorough and exhaustive research of Tom Woodhouse, a key member of the team during the project, led to the discovery of original vendors, materials and components needed in the journey to return the Spitfire to a pristine 1940

factory-fresh standard. PPS, like other warbird rebuilders, have had their share of motion picture and television credits on films such as *The Mummy*, *Indiana Jones and the Last Crusade*, *Battle of Britain*, *High Road to China*, and the tv series *Piece of Cake*.

Across the Atlantic, Americans are deeply and passionately involved in the warbird movement. Steve Hinton is one of the major players through his roles as president of Fighter Rebuilders at Chino, California, and his involve-ment as a warbird pilot *extraordinaire*, movie pilot, unlimited air racing champion and air show performer. He is renowned for being willing and able to fly almost anything. Steve is also president of the Planes of Fame Air Museum at Chino's Cal-Aero Field. He grew up around war-birds, working as a volunteer with Ed Maloney's Planes of Fame collection. After a ride as a teenager in the muse-um's AT-6 trainer he was hooked. His chance came when contractor, Leroy Penhall, moved his company to Chino and employed Hinton to take care of his immaculate P-51 Mustang fighter in exchange for being allowed to fly it. Penhall also owned an F-86 Sabre of Korean War fame and, when Steve gained his instrument rating, Penhall checked him out in the jet. Possessed of superb flying, mechani-cal and business skills, along with a winning personality, Steve was well placed to succeed when he started Fighter Rebuilders at the Chino base. The work began with the restoration of a Curtiss P-40 for Flying Tiger Airlines and gathered momentum as major collectors, including Stephen Grey in England and Bob Pond in California, began sending their proj-ects to Hinton. He does restoration work for other clients such as California collector Tom Friedkin and for the Chino museum. He says: "We'll do other projects but we don't advertise for it."

Of the many fighter types Steve has worked on, the Spitfire lingers in his memory: "If you get past the weird British stuff, it's really a beautiful air-plane. It's different . . . instead of hav-ing a lock nut, it will have a nut on a bolt and then pound the threads over so it won't come off. Their cotter pins are the stingiest things ever. And they use a pneumatic system that we didn't deal with back then. My point is, your first impression of a foreign airplane is that it's a piece of crap. 'Look at this! Why did they do it that way?' But when you finally under-stand how it works or why they did it that way, all of a sudden you think, 'Oh yeah! Take the gear handle, for example. It turns out it's not just a gear handle. It's the one valve that controls the entire system. It's just a different mentality, but it works well." Steve's aim with Fighter Rebuilders is to tailor the work to the workforce he has. "We're not just picking up every project that comes around; it's not that kind of business at all. Our main goal is in support of the Air Museum. We're caretakers right now. Hopefully, this will last longer than my life and our current group here. We've got to make sure these airplanes and the veterans' sto-ries are out there forever. That's the whole magic of these airplanes. They inspire. I'm lucky to be a part of it."

In January 1957, Ed Maloney opened his Air Museum at Claremont, California with six aircraft he had managed to save from the grasp of scrap dealers eager to cut them up for profit. 300,000 military planes had been produced in the United States during WW2 and most of the surviving examples were stored in aeroplane graveyards in California, Arizona, Arkansas and Oklahoma after the war. They were eventually sold to recover their scrap value. Maloney continued to collect WW2 aircraft types and, as the collection grew, moved the Museum, first to Ontario Airport, California, and later to the Chino field which, during the

the war, had been the site of the Cal-Aero Flight Academy, a training school for Army Air Corps cadets learning to fly some of the aircraft now preserved by the Museum. The collection features the only totally authentic flying example of a Japanese Mitsubishi A6M Zero fighter plane, which even retains its original engine. In addition to monthly mini-air shows, Planes of Fame presents a major annual air show.

One of the newer and highly promising warbird collections is that of Microsoft co-founder and multi-billionaire Paul G. Allen. The mission of his Flying Heritage Collection is to collect, restore, fly and preserve combat aircraft and artifacts representing technological, ideological, political, and economic views of aerial conflict in the 20th century, with emphasis on World War II and the Cold War era

Allen: "There is an elegance to the planes. That was an era when the technology leaped ahead. That was a watershed period."

Of particular interest is the Messerschmitt Me 262 *Schwalbe* jet fighter that the FHC bought from the Ed Maloney's Planes of Fame Museum for restoration to airworthiness. The notoriously difficult and unreliable Jumo 004 engines of the aircraft are

AN ENGINE RUN TEST OF A NEWLY-
RESTORED HAWKER HURRICANE AT
THE FIGHTER COLLECTION FACILITY,
DUXFORD, CAMBRIDGESHIRE.

being rebuilt with new, custom-made turbine and compressor discs, blades, stators, and new "hot-section" parts made using current alloys. The Allen team is looking for a new service life for the rebuilt engines in excess of 200 hours, a substantial improvement on the ten-hour maximum service life of the original Jumos. More than a dozen FHC acquisitions currently await restoration. They include a CASA

2.111 (Spanish-built Heinkel He-111 bomber), a North American F-86A Sabre jet fighter, a De Havilland DH 98 Mosquito, a MiG-21 jet fighter, a BAE/Hawker GR-3 Harrier jump jet, and a Messerschmitt Me 163 Komet rocket fighter. Allen moved the FHC from its Arlington, Washington location to a new facility at Paine Field, south of Everett, Washington in 2008—an airfield it will share with Boeing's

Future of Flight Museum.

One of the most intriguing warbird ventures of recent years is that of a small group known as Classic Fighter Industries, in Everett, Washington. Aeronautical engineer Steve Snyder led the group and subcontracted Herb Tischler's Texas Airplane Factory, of Fort Worth, to restore an original 262 and build five new examples. After a difference of opinion between Snyder

and Tischler, Snyder brought in Bob Hammer, a Seattle-based Boeing executive, moving the project to the Pacific Northwest in 1999. Later that year Steve Snyder was killed in the crash of an F-86 Sabre jet. With his loss, the owners of the first two Me 262 project aeroplanes decided to finance the effort.

The Me 262 German jet became operational near the end of WW2 against the giant bomber streams of the American Eighth Air Force in their attacks on Nazi industrial targets. Joining the action too late in that conflict to affect the outcome, the twin-engine fighter was capable of 540 mph in level flight, far in excess of its propeller-driven adversaries. More than 1,400 were built and of those there are eight known survivors.

Of the five new-build 262s, two are designed as tandem two-seaters and two others can be converted from one to two-seats. The most significant changes from the original Me 262 are the use of General Electric J-85 jet engines instead of the Jumo 004s; reinforced landing gear, a modified throttle assembly, and additional braking capability.

After many years of work, the first of the new 262 jets lifted from a runway at Paine Field, Washington, on December 20th 2002. At the controls was Wolf Czaia, a retired American Airlines 757/767 captain who had formerly flown F-104 Starfighters in the modern German Luftwaffe. After the flight, Czaia reported: "A pleasure to fly. Overall, a great first flight."

The Commemorative Air Force was formed (as The Confederate Air Force) in 1957 when five friends in Texas began acquiring surplus WW2 fight-

LEFT: A BRISTOL BLENHEIM BOMBER RESTORATION; BELOW: THE LANCASTER BOMBER OF THE BATTLE OF BRITAIN MEMORIAL FLIGHT, CONINGSBY.

ers. They set out to preserve, in flying condition, as many individual warbird types as they could, and they built up a large organisation with nearly 150 aircraft and seventy regional groups in twenty-seven states. Their smallest plane is a Stinson L-5 Sentinel; their largest a Boeing B-29 bomber, the only flying example in existence. With more than 11,000 members ranging in age from eighteen to 100, the CAF is now headquartered at Midland, Texas, where it also operates the American Airpower Heritage Museum, dedicated to helping preserve the

WW2 American aviation heritage. Clearly, restoration and maintenance is a mammoth task in the organisation, requiring the constant participation of hundreds of volunteers as well as very deep pockets. The pilots and aircraft of the CAF perform annually in air shows before 10,000,000 people in the United States.

The Military Aircraft Restoration Corporation, the warbird collection of David Tallichet, was founded in the 1970s as Yesterday's Air Force. It is remarkable for a long-running involvement in the salvage and recov-

ery of WW2 aircraft—many of them from the Pacific and Papua New Guinea in particular. One of the largest aircraft wreck recoveries took place between 1973-75 when an A-20, three P-39Ns, four P-39Qs, six P-40Ns, and four Beaufort torpedo bombers were all recovered from Tadji in Papua for eventual restoration by the MARC people. David Tallichet is chief executive of Specialty Restaurants Corporation which includes a number of military aircraft-themed restaurants in the U.S. He flew twenty-two missions in WW2 as a B-17 pilot with the

100th Bomb Group of the Eighth USAAF in England.

Another WW2 aviator who has become one of the world's most important military aircraft collectors is businessman and philanthropist Robert J. "Bob" Pond. His Palm Springs Air Museum at the Palm Springs, California, Regional Airport is home to his legendary collection of World War II aircraft, as well as his automobile and aviation artifacts collections. His commitment to the preservation of the planes, to maintaining and flying them in honour of the sacrifices of his fellow wartime fliers, and to furthering education about the war, is unsurpassed. The museum is a non-profit educational institution whose aim is to exhibit, educate and demonstrate the role of the World War Two combat aircraft and the role that both pilots and American civilians had in achieving victory. As the museum's website states: "The significance of World War Two is unparalleled in all of history; it was the greatest, most costly conflict ever fought, taking the lives of more than 70 million people. It was air power that altered the outcome of that war and forever changed the lives of every person alive today."

Bob Pond trained to be a naval aviator and learned about aircraft carrier flying in Florida during 1944. While in the Navy he took advantage of opportunities to fly a range of bombers and multi-engine aircraft, including the Consolidated Catalina and the naval version of the B-24. After the war, he went into the family business, Advanced Machine Company, which made floor scrubbers and polishers as well as vacuum cleaners. His love of flying persisted, however, and, after more than sixty years, his logbooks

AR213 AT WYCOMBE AIR PARK (RAF BOOKER) IN THE 1970S.

record an astonishing 22,000+ hours of flight time.

Pond began collecting warbirds in 1970 when he happened to meet a retired Northwest Airlines pilot who wanted to sell the P-40 Warhawk and P-51 Mustang he had rebuilt. Bob bought the fighters and was checked out in them by the airline captain. The planes are still a part of the museum today—the Mustang being Bob's personal favourite to fly.

His other collecting passion is old cars and his superb examples include Aston Martins, Rolls-Royces, a 1948 Tucker and a 1935 Chrysler Airflow.

Over the years, Pond's fascination with warbirds continues, as does his wish that as many of the museum's airworthy planes as possible are flown every Saturday in the ideal desert climate, from October through May, for the pleasure and education of visitors.

To those in charge of the museum, its most important aspect is the living history and eyewitness accounts of the hundreds of docents who volunteer their time to explain the exhibits. Many of them served and survived in air combat during World War Two. They are considered a priceless resource and asset of the institution.

Of approximately 300 air museums in the United States, the Palm Springs Air Museum is one of only a dozen that continues to fly its aircraft after the heavy increases in fuel and insurance costs brought about by the terrorist attacks of September 11th 2001. As Bob Pond has gradually turned over the reins of his collections to others, he reflects: "The museum is really the centre of my life now. We have the most wonderful group running the museum and it's great to sit back and enjoy the directions

C-47 SKYTRAINS AND A B-17
BOMBER AT DUXFORD.

REBORN

Personal Plane Services' Tony Bianchi: "When the matter of doing the first complete restoration of AR213 came up, I had already discussed with the owners, the fact that the aeroplane was showing some signs of needing attention. I knew that they would want to go for a pristine, 1940 standard and would want the aeroplane done exactly as it was in its day. We had agreed that the way to proceed with the restoration was to get the aeroplane absolutely 100% to that standard. Before we even started putting estimates in for what it would cost to do it, we pretty much had it in mind that it had to be done that way; there really was no other way of doing it.

" We were then doing aeroplanes for Kermit Weeks and he is another perfectionist. His aeroplanes had to look and smell exactly as they would have done in their day, or as best as we know it would have done. Kermit was really the first guy to do aeroplanes that way. So, we were already doing restorations of that kind; no one else in the UK really was. In some cases, they were taking all the original equipment out, including radios, armament, instrumentation, gun sights, and a lot of unique Vickers-Supermarine fittings and throwing it all away. Some people were trying to put that stuff back in their aeroplanes, but didn't have it. We were going around rescuing the stuff that people were throwing away on the rubbish tips. We were doing some restorations for other clients who just wanted masses of radios and modern equipment in them, changing this and that. We were doing our best with Kermit to acquire aeroplanes and turn them back to the way they were. I said to him, 'Look, the right way to do these

things is to do them exactly the way they did in their day'. In response, his inference was 'I agree. If it doesn't cost any more, why not do it the correct way?', and these were some pretty expensive things. As an example, he wanted to make some hood runners for his Grumman Duck. They were a kind of a figure-of-eight and needed folding on a press brake and he couldn't get anyone to do them. Kermit is a very clever and competent engineer himself; he can design things and he knows how to make things work. He'll just spend his time doing things properly and he got a lot of good machinery when he was down in Miami. He has a fantastic machine shop having bought all kinds of machines from various airlines that had gone bust, so he could make pretty much everything and he wanted to make these hood runners. However, they were about twelve feet long and he eventually found that there was a company in Miami, a machine tool supply shop, doing a very expensive press brake, and he bought it to make two hood runners, on the basis that he was going to use it again. That's what a perfectionist he is. In the case of his Tempest V, I said to him, 'This is the way they made these bits on the Tempest' and he'd say, 'Do it. Make it as they did.'

"There are short-cuts you can do on these aeroplanes. Nobody had ever rebuilt a Tempest V because there are only two in the world, his and the one in the Hendon Museum and we had to borrow parts off the Hendon Museum aeroplane, from the bulkhead forward, and make everything. For the cowlings, it took us over two years to make the jigs before we ever cut any metal for them. It was a huge job of

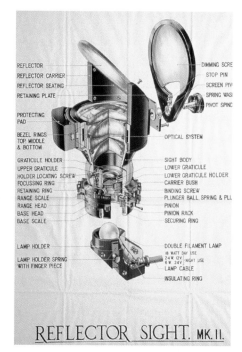

REFLECTOR   SIGHT.   MK. II.

REFLECTOR
REFLECTOR CARRIER
REFLECTOR SEATING
RETAINING PLATE

PROTECTING
PAD

BEZEL RINGS
TOP, MIDDLE
& BOTTOM

GRATICULE HOLDER
UPPER GRATICULE
HOLDER LOCATING SCREW
FOCUSSING RING
RETAINING RING
RANGE SCALE
RANGE HEAD
BASE HEAD
BASE SCALE

LAMP HOLDER

LAMP HOLDER SPRING
WITH FINGER PIECE

DIMMING SCRE
STOP PIN
SCREEN PIV
SPRING WAS
PIVOT SPIND

OPTICAL SYSTEM

SIGHT BODY
LOWER GRATICULE
LOWER GRATICULE HOLDER
CARRIER BUSH
BINDING SCREW
PLUNGER BALL, SPRING & PLU
PINION
PINION RACK
SECURING RING

DOUBLE FILAMENT LAMP
18 WATT DAY USE
24 W 12V
6 W 24V NIGHT USE
LAMP CABLE
INSULATING RING

PRECEDING SPREAD: AR213 OVER OXFORDSHIRE; THIS PAGE: IMAGES OF AR213 DURING RESTORATION.

work, but Kermit being Kermit, it had to be done the way they did it in its day. And the Tempest is a highly complex aeroplane—far more complex than the Spitfire. Every piece of aluminium is heat-treated on an aeroplane like that, so even the cowlings are not just ordinary pieces of material. They go straight into an oven and get heat-treated. You've got to work the metal first, then heat-treat it, then cure all the problems of changing shape once you've heat-treated it. We got into some quite high-quality manufacturing philosophy within the company, so, doing a Spitfire to the sort of ultimate level of restoration and quality was not a problem. The problem lay in making it look like it hadn't been done . . . like it was a fresh aeroplane that's been on a squadron for ten hours and it's got a little bit of patina in it and is just aging down gently. That's the philosophy I've tried to put onto the aeroplane. We did that with a lot of Kermit's stuff. Some of the aeroplanes, of course, when he'd used them for a hundred hours over a period of five years, they'd start to look knocked around the edges like a squadron aeroplane. Tony Bird is excellent at doing that; he knows how to age things nicely. He'll make something and make it look like it's fifty years old as he's doing it.

"We look at other people's Spitfires, other people's restorations, and they miss detail and passion. They are either incredibly 'brand new and polished looking', or they've gone halfway to the way we do it, but have missed something out. I look at it and think 'that doesn't look like a squadron aeroplane; that doesn't look like an aeroplane that's come from the factory. It looks like something that somebody has rebuilt who doesn't know what an aeroplane should look like.' Even the sort of '30s aeroplanes that we have done for Kermit, we had to get them looking and smelling right. One of the dangers that goes with it, you can do it to too much of a degree and ten years down the road the aeroplane looks like it needs restoring again because you've almost gone too far with it. It's largely cosmetic but no matter what, it has to be properly airworthy.

"We have compromised ourselves in some cases by going the way they did it in World War One and the post-World War One era, doing the paint finishes that they did and we'd have damaged fabric now. Of course, because they are natural materials, not man-made, they suffer from ultra violet light damage, and the UV light is worse now than it used to be. You cover the control surfaces in Irish linen and once you've got it all doped you then put acres of silver paint on it to reflect the UV light. In the case of Kermit's World War One aeroplanes, we did exactly what they did in that war. We mixed aluminium powder with the paint so you'd get a slight glistening effect, and that's why some people thought they were glossy. You hand-painted everything on. They were all hand-painted. Nobody sprayed much in those days, especially on the squadrons. And we've had this fabulous finish which everybody says, 'God, is that what it looked like?' Yes, that's what they looked like, but within five years the UV light's killed it all because it hadn't been protected properly. Some areas had got enough aluminium powder in them to reflect the sun and other areas hadn't, so you could put your finger through the fabric in some places. We compromised ourselves but we had done it the way it should be. We knew that we'd done it correctly.

"Our restorations, the way we do them, are not everybody's cup of tea; there are people out there from the warbird world who don't like what we do because they think you should have it all glistening like *Grand Warbird Winner, Oshkosh*, glitzed up to the nines. A lot of people want to see that. In my mind, Kermit is responsible, anywhere in the world with these restorations, [the kind that we do] for 'let's make it look like and smell like it really was; it's got to be the real thing. What's the point of doing it unless it's just as they did it?' He's the architect of these things. He got us all into it.

"I've got a Fokker E3 Eindekker replica. Kermit's flown it and he's always wanted to build 'an original' and there are no original E3s in the world, other than the one in the Science Museum in London. They were tricky to fly because they had an all-flying tailplane and warping wings. Some of them were quite high-powered aeroplanes. So, we started on this folly to see if we could build, say, three Eindekkers. I wanted a 'real one' anyway . . . to put an Oberursel engine in it, do it all properly. We started looking into it, we got what drawings existed and started to go through it. It had all metric-sized tubes which decrease in diameter as they go to the rear of the fuselage and it's an all-welded structure. Antony Fokker designed the first electric welding plant; as far as we know, he was the inventor of electric stick-welding. He was the first guy to have all steel-tube welded fuselages. All the

Fokker fighters were welded steel-tube. So, Kermit said, 'Right. We've got to do it the same way, but let's build Fokkers using arc welding plant. Let's start off the way Fokker did it. Let's get the drawings for the arc welding machine, make the arc welder first and then stick-weld everything.' We never did do it in the end because it was typically expensive and Kermit had gone off on another track. Perhaps we'll return to it one day. But his philosophy is why that aeroplane [the Mk Ia, Spitfire AR213] is like it is. It has paid off for Personal Plane Services. It is now the company philosophy. We always wanted to do it, but the opportunity to perform on Kermit's aeroplanes gave us the opportunity to perform on other people's aeroplanes and they benefit from it.

"Every other Spitfire we see is not done quite correctly, by comparison. We know that. We're not criticising other people's restorations; it's just a different philosophy. We know that the effort that Tony and Tom and Frank put into doing the fabric work on the Mk I Spitfire, we know it is unlikely anybody else does the same. We've got other Spitfire rebuilds around us that have been done by other companies and you've only got to tear the fabric open and look inside; they are not done in the complicated, old-fashioned way that Vickers-Supermarine employed. The Mk I is done that way. In a lot of cases, other companies have had to cut corners because they were on a tight budget from the owners and, equally, there were cheaper alternative ways to do it. We've been instructed by clients that they don't want an original instrument panel. They want all radios bristling from everywhere in order to be able to get

around in pretty much any weather throughout Europe. The Mk I Spitfire is going to be flown sensibly, so we are unlikely to be trekking around Europe in poor weather. It's paid off for me because I have the satisfaction now of knowing that we have probably done the ultimate early Spitfire.

"Another very correct Spitfire was Kermit Weeks' Mk XVI. The Mk XVI is not the nicest aeroplane to do it with, but we did go pretty deeply into it. It had a lot of original equipment in it. Effectively, the guns and armament worked and it had all the original armour plating, radios, oxygen system. We went as far as you could go, sensibly. It was a two-year restoration and it did form the basis for doing this Mk I aeroplane.

"Out of a hundred guys who want these things done, there are only one or two who want it done properly. That's not to denigrate what other warbird owners and rebuilders are doing. They're doing it their way. It's just that I feel that that aeroplane [AR213] will stand the test of time for our having gone the route that we have. It's bloody expensive to do it this way and actually, it's a difficult philosophy because I find that other customers come to us and want things done and I have to say, 'Look, it's not the right way of doing it. I will do it, but there must be a compromise somewhere along the line.' Fortunately, I'm winning people round. I've just sold our Yak-11 to a client of ours and we're going to partially rebuild it in Hungary. They've got the capability there because they are working on Russian/Soviet-type aeroplanes all the time, so you know all the correct nuts and bolts and switches and things will be used . . .

The philosophy is 'get it absolutely a hundred per cent, because you stick the value way up if you do that.' These days, everybody changes things. I've flown Spitfires that have been dramatically changed. I flew a Mk XVI in the States that had done the rounds; been over here, was rebuilt in Chicago back in the early '60s, came over here in the '70s and then went back to America. I was flying it down in Arizona and it had had the most awful things done to it. A guy rebuilding it in the '60s didn't know it had a thermo-pneumatic valve for opening the radiator flaps, which was completely automatic and trouble-free. And he put a Cessna 150 wing-flap motor on it with a switch, 'Why do those Brits do these things with all the pneumatics and temperature-sensitive bulbs that go in coolant pipes? Aaah, just get rid of all that. Put a switch there and a little electric motor that drives a rad flap up and down. Simple.' And he's right, but it's not correct. We've seen them here, one that was rebuilt in this country and the guy changed everything around, changed all the hydraulics around; made the hydraulics work things that it shouldn't work and the air wasn't working when it should work. It was unflyable."

In his special approach to the restoration of Spitfire AR213, Tony Bianchi directed the efforts of a uniquely talented team: "I knew the owner would want his Spitfire done as perfectly as possible, especially as AR213 was an outstanding example of an early Spitfire that had just been kept nicely serviceable and not really changed. People who used to know the aeroplane back in the old days would look in it and say it hadn't changed. It's

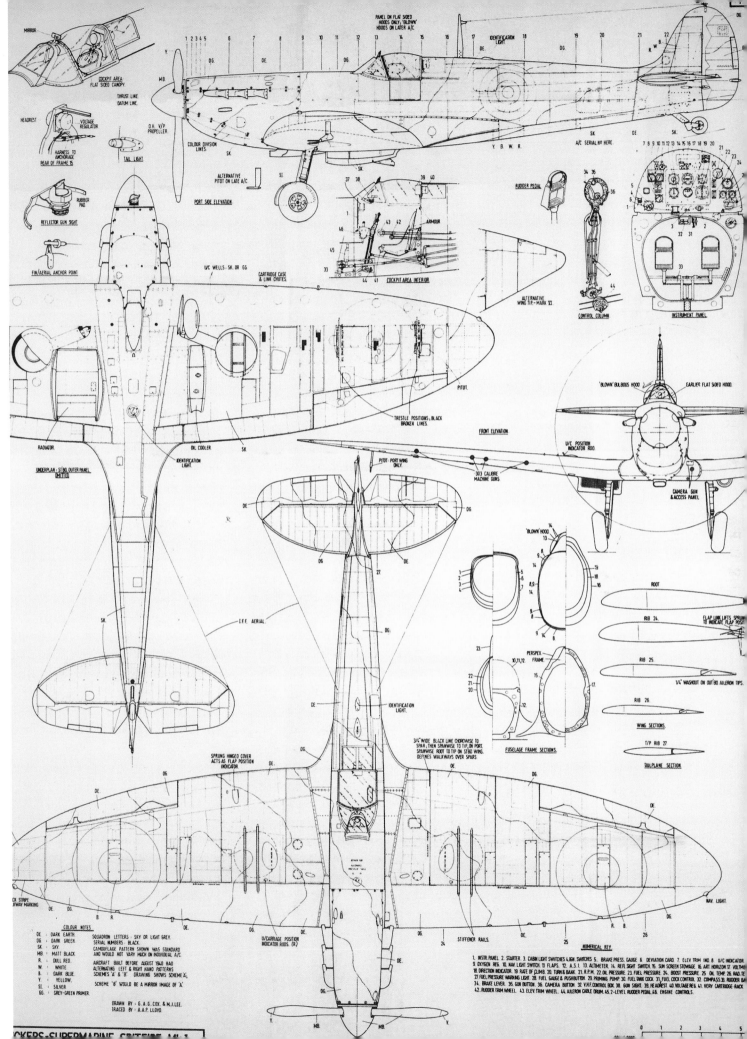

VICKERS-SUPERMARINE SPITFIRE MkI

about the same as it was in wartime. Our goal was to try and preserve that and maybe make it look a bit tidier because it's going back to being 'a nice, worn sports jacket', which is the effect that I knew we all would want. We've gone into it much further. Tom Woodhouse has been responsible for obtaining everything including originality and the chasing down of original equipment that we were missing, history and documentation etc. The research and tracing of originality is a speciality of Tom's. Franco Tambascia is responsible for the engineering supervision of everything and I oversee him as the central point from customer to engineering, fronting it and trying to control the pennies to a degree. Unfortunately, with this sort of one-off project, it's very difficult to control them because things suddenly start hitting you from all angles and then you realize you've actually got to spend a great deal more of someone else's money than you wanted to.

"In the restoration work itself, we've gone a step further. Even in the most minor details, like the rear seat back and headrest, for example, which was probably not the original that came in the aeroplane, it was a dark brown and very badly worn, with all the stuffing coming out of it. We thought we would re-do it and Tom found some samples of very dark green leather that had gone on the rear headrest/crash pad, which we'd been trying to source. So the seat back and crash pad were done in that particular unusual specification of leather and, as original, filled with horse hair. We really went into the detail of what and where everything was positioned in the cockpit—as it would have properly been when the aeroplane left

the factory. Hopefully, it will settle into having a nice, warm, 'used' feeling to it, but correct.

"As far as modern additions that we have had to put in go, the Civil Aviation Authority in Britain still require a NO SMOKING notice and a starter engagement warning light, which we've incorporated into the normal Spitfire-pattern generator failure or fuel pressure warning lights. We've just made the additional warning light so it doesn't look 'out of period.' The CAA require a bit of cockpit placarding, but nearly everything else is as it was. We don't use the emergency air bottle for getting the undercarriage down, as they did in the period. That was a small bottle with a short lever and a trigger on the top of it. It had a needle that just pierced the bottle and allowed a thousand pounds of compressed air into the hydraulic system, which put the undercarriage down once you'd got it unlocked. That bottle is very difficult to get recharged because they have to recharge it and then put a lead or copper cap in the end of it. They were pretty unreliable anyway. We now have a rechargeable bottle which sits behind the seat. You can't see it that easily. That's a modern addition, but the only one we are doing, other than the radios. We have a radio and a transponder that fit in the original map case, which you can't easily see.

"The restoration process has gone relatively smoothly, but not entirely so. Before our involvement, some repairs had been done in the wartime period or during the *Battle of Britain* film, and it was a mess. In the case of restoring the fabric-covered control surfaces, to completely dismantle and

strip the very fine tubing in the elevator and the rudder was a good couple of months work for one guy. We had to wait for work from sub-contractors who have been overloaded with work, and that slowed up the process. There aren't many people we trust to do this work. We were a bit hampered by the delays and got to the point with some things on the aeroplane where we had to stop until we got more spares to continue. For a long time we didn't have an engine mount, an engine, and a propeller. Therefore, we couldn't finish the cowlings. We had to make a new top cowling, the old one was in such bad condition, having changed shape quite a lot and been repaired too many times. There were a multitude of things from the firewall forward that we couldn't touch without the engine installed. Our electrician couldn't finish his work because all the wires that stick out of the bulkhead going up to the engine couldn't be cut or trimmed off. We didn't have the wings on for a long time so he couldn't finish all the electrics in the wing-to-fuselage hook-ups. We were waiting for coolant pipes. Some of them needed to be installed and set up in the wing before the wing went on. For a long time we were in the difficult position where we were waiting for everything to arrive and then we would need all hands to put it together.

"As to the future of the aeroplane, we are all fairly united that it wants flying regularly and to be seen around, but sensibly and practically. It doesn't want over-using or over-exposing. It would be a pity to join the air show circuit and have the guts thrashed out of it. One of the risks these days is that you get sucked into all these formation things and the aeroplane gets lost in a big formation of other Spitfires and warbirds. I think it needs selective show situations where it is on its own, taken as a bit of artwork on its own, rather than being just another Spitfire among fifteen aircraft all flying around in formation. The risks are a bit higher doing that as well. We agree with the notion to let as many people, old and young, get pleasure out of seeing a unique Spitfire being nicely shown and presented.

"Initially, Jonathon Whaley and I are going to do the flying on it. Carolyn Grace [who owns and flies a two-seat Spitfire] has flown it and she should definitely be keeping herself up to speed with it. She has never flown another single-seat Spitfire. Once, when Victor Gauntlett was around, he wanted a go in a two-seat Spitfire and wanted his family to fly in it. He asked, 'Can you get Carolyn to come along? I'm sure she'd like to have a go in the Mk I.' I said, 'Yeah, I'm sure she would.' I rang her and she said, 'I'd love to, but why would you want to do that? Nobody will let me fly their aeroplanes. Maybe it's because I'm a woman.' I got on to Victor and told him what she had said. He said, 'That's absolutely bloody atrocious. You get her down there and get her in the aeroplane and send her off. I'm not bloody well having that.'

"Carolyn flies the Spitfire 100% by the book. She won't mess around. What it says in the book, she flies it 'textbook'. She has the respect for the aeroplane because she owns one and she knows what it costs if you have a problem. I think one of the main things is that people should be aeroplane owners before they fly other people's aeroplanes, so they have

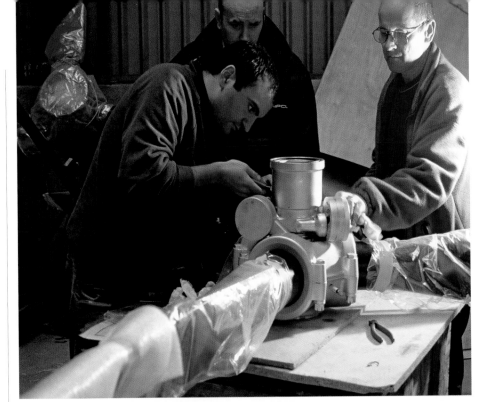

BOTH: FRANCO TAMBASCIA AND TONY BIRD DURING THE RESTORATION OF AR213.

respect for them. She's probably got 500 hours of successful Spitfire operation. She takes people flying and she's very responsible and does exactly what it says in the book. She doesn't step outside those areas. She flew the Mk I 'textbook' the first day she flew it. Fortunately, most of the people who have flown it did so that way. A few people did get into it in earlier days who really hadn't got the experience and, fortunately, they did very sensible, gentle flying, but in some ways they shouldn't have been flying it. It was flown during the '70s, '80s and '90s with great care and respect, and that doesn't happen with warbirds as much these days. New people are jumping in and out of these things who really don't know what they are doing. Some of them do, and they are very good, but a lot of people don't really understand how all the systems

work; they just kind of hope nothing goes amiss with it. Do you gain anything by letting people like that fly it? You don't."

Tom Woodhouse of the highly-skilled team that restored AR213: "We started taking it apart in 2001. The tailplane, but not the control surfaces, both ailerons and the flaps had already been restored. It was basically the fuselage, wings, engine bearer, cowlings, and of course, making all the other parts. It had to be a bare-bones restoration; we had to pull everything completely apart. Sixty years of wear and tear, everything had to be looked at. One of the things we found was that the starboard wing was not as well-manufactured as the port wing, which was really nicely done. The starboard wing looked like it had been done in a severe rush. Certain things

looked cobbled together because they didn't really fit. They'd just drilled it off and whacked it in. We replaced all of the rivets, every single one. On the wings, we were able to keep at least 60% of the skins. The starboard wing needed more new skins than the port, which was reasonably good. I would say that at least 70% of the fuselage skins are original. But the aeroplane has been repaired in the past, so some of the skins aren't original from the factory anyway. A lot of that work was done for the *Battle of Britain* film.

"We mainly had five people working on the restoration. There have been at least three or four companies doing sub-contracting work as well, the main one being Supermarine up at Stoke-on-Trent.

"The propeller came from a Canadian-built Fairey Battle which was purchased by the builder Charles

FAR LEFT: AN ENGINE TEST FOR THE MERLIN OF AR213; LEFT: THE REASSEMBLY PROCESS.

Church in the 1980s. It was removed from the aircraft and intermingled with Church's other Spitfire spares. After Church was killed in the crash of his Mk V Spitfire, Kermit Weeks bought the other Charles Church Spitfires and all the spares. Kermit had no need for the Battle prop and sold it. We later acquired it and when we undertook the restoration of AR213, we had a company in Germany re-profile it to be exactly the same as the original Mk I De Havilland DH5/29 three-blade propeller.

"During the Second World War, the De Havilland Spitfire propellers were sent out with long blades which were then cropped and re-profiled when used for fighter application. They were all made to a set length and were then cropped to suit the aircraft they were going to be used on. The Battle has a slightly longer prop because it had a

higher undercarriage. The Spit has a slightly shorter undercarriage so the prop had to have slightly less diameter for ground clearance."

Tony Bianchi: "When I think back about this Spitfire, some of the 'rough rides' I've had in it from time to time in my early days of flying it . . . some of the things I would never have got away with in other aeroplanes . . . a lot of my philosophy to me means that in many ways the Mk I is actually a great saviour. It's such a good aero-plane to fly and such a delightful, safe seat, that you probably push your luck too hard with the environment. I think, when I was younger I did, when I look back to swanning around in bad weather, looking at fuel gauges and thinking, 'Oh, I've got enough left.' I had half a dozen occasions where I had near-scares with it, that

were offset by several hundred hours of superb flying where it was an utter delight to fly and operate. You get a gorgeous summer evening when you are thundering across northern France, coming back from somewhere and everything is working nicely and you've got the canopy slid open and you think 'this is absolutely perfection' and you couldn't wish for a better aeroplane. That's one of the things that a lot of people probably don't think of these days. There aren't many Spitfire owners who get to appreciate the aeroplane they own. They don't appre-ciate owning and getting used to just one aeroplane . . . how people must have felt during the war, flying noth-ing else but early marks of Spitfires for two or three years. They really got themselves into the aeroplane. Some of these people flew Spitfires all the way through the war, on flight testing development work, and a bit on operational squadrons. People like Jeffrey Quill, who were flying them up until the 1960s—those guys must have absolutely got right down to fine limits, flying those aeroplanes and utterly trusting them. It's all I used to fly at one time. Then I started flying other aeroplanes and used to get in them and think, 'This is all right, but it's a dog compared with a Spitfire.' The Mk I, in particular, is a very overpowered light aeroplane, but it's actually quite docile to fly. You can get complacent with a Spitfire and then get 'a wake-up call'. You get away with it and then you don't do that again. It's a friendly old thing. Certainly, a Mk IX is not as friendly an aeroplane.

"When we did Kermit Weeks' Mk XVI, which is only 'a tear-drop Mk IX', we put everything on it. We had all the armour-plating. We had armour-plating on the seat; armour-plating around the engine; armour-plating around the fuel tank. We had twenty-millimetre cannons in it. We had half-inch machine-guns. We had ammunition. We had old radios; all the original kit that you get in it. It became a more tricky aeroplane and not nearly as pleasant as a standard, stripped-out Mk IX variant, which is mostly what people are producing after restoration these days. I thought it felt more like flying a P-51 around, because it was getting closer to P-51 weight with the same engine. It used to go out of here like it really didn't want to go anywhere. It didn't fly as well as a two-seater, which is pretty much a heavy variant. Interestingly, if you start out at, say, five-thousand feet

in a Spitfire and push it over into a dive, it will normally hit pretty high speeds on the way down, without needing to use much trim change. But that bloody Mk XVI, if you were flying it around, the moment you rolled into a turn, you were trimming the whole time to keep it in the turn. If you rolled out you'd then have to trim it again to get it straight and level. To dive it to VNE, maximum dive speed, it required ten-thousand feet and constant use of the trim. It could be a pig to fly, but it was absolutely correct and, no doubt, you could get used to it and enjoy it.

"We had a very good friend here who had flown Spitfires, Tempests, Typhoons, and P-51s, all marks, Mosquitos, everything. He worked here for us for a time and he always said to me, 'you know, I actually prefer

AS THE RESTORATION NEARS COMPLETION, ENGINE TEST RUNS CONTINUE IN PREPARATION FOR FLIGHT.

a P-51 to a Spitfire, always.' It wasn't until I flew a Spitfire which was all fully-loaded that I thought, 'yeah, I'd prefer a P-51 . . . to be fighting with, for low-level, for a bit of dogfighting, a good all-rounder with range . . . give me a P-51, whereas the Mk I, it's difficult to set that up to those weights. You've got the nose which is probably eighteen inches shorter, a slightly lighter propeller, a lighter spinner. You've got eight machine-guns instead of two bloody great cannons, and we made the cannons out of aluminium, but you still couldn't carry them. It was SO heavy and with its ammunition and the drum belt-feed and all the mountings for it, it was a huge weight. We just copied everything in aluminium. The radios are bigger and heavier; everything you put in it is bigger and heavier. It had two bloody great radiators, an intercooler and a big oil cooler, all these bits and pieces that you add to it, you can't do it with a Mk I. The Mk I will go up a small percentage in weight. We're gonna be heavier, but I'm sure it will still be a nice flying aeroplane. It won't be absolutely as nice, but if we pull all the guns out of it, which we don't have to fly with, and that's only a twenty-minute job, and then it will be back to the way it was. It might even fly slightly better in some ways because we'll get the centre of gravity to where the book figures were. All this stuff about, 'Oh, the Mk IX is the best flying Spitfire in existence'—don't believe it. It's nice, but it's not *that* nice.

"Flying the Mk I in the future. . . regular use; don't expose it to anything that's too risky. If the jobs come up for it with the odd air shows, film work

and that sort of thing—providing we're not asked to do anything that's ridiculous, I think we just operate it as it was always operated. But now we've got some confidence that everything's been looked at and there's nothing that's going to let us down . . . there's always something that's going to be letting us down somewhere along the line because we are still dealing with bits and pieces that are sixty years old, so there's always going to be issues, but we have a more general confidence in that it's all been looked at and overhauled. We'll press on with the same philosophy. I think we'll just do everything we can with it. There is going to be enough work in carefully operating it, to be honest, and being selective in how it's operated, carefully operated . . . there's gonna be some work to do there.

"To get the CAA to be happy with the aeroplane, we've got to be happy with it first, after a very in-depth inspection. For instance, take the port wing. If there is something that has still got to be done on it, you can't sign the work off. All the work sheets are there; the inspection schedule is there and it can be inspected to a point, but if all the work isn't finished you can't sign the wing off. We had issues with everything during the last stages of the restoration, bits we were waiting for, bits we hadn't solved a problem with, that we couldn't solve because we were waiting for parts to come back or waiting for someone to involve themselves in a particular area to try and sort out. We were struggling to get all those bits and pieces together. Mainly it was engine installation. Once we finished the front end we could start moving back, hood and windscreen, controls, some of the

wiring, then out onto the wings. We had an undercarriage problem. As we were able to put the aeroplane together, we found out what the snags were. The snags never bothered us when we were flying the aeroplane before the restoration because we just kept it serviceable. You took things off and saw that there were no cracks in there and put them back on. But nearing the end of the restoration, for example, we found that we had a huge amount of wear in one undercarriage leg, the pintel that it rotates on. Well, it flew like that; they'd accepted it, but we couldn't have it flying like that now because we were rebuilding the aeroplane, so the leg had to come off and we had to start making some fairly major items for it. That delayed us a lot. We did retractions with it but we still had some cable problems in the retraction rotating of the pins, to get the pins out.

"Everything we tried to fit had never fitted properly before. It had been slung on for the *Battle of Britain* film and accepted that it was all right. We never had the oil coolers off the aeroplane before. We came to hang the oil coolers back on and they were touching everywhere. They've had cracks in them before which were repaired and we'd seen why they had cracks . . . because bits of skin were chattering away at them. When Tony went to fit an oil cooler on the aeroplane he found that one rib was getting in the way of the oil cooler and was touching the fitting that attaches it. Part of that rib had to be taken out and re-rivetted in a different way, which meant taking the oil cooler out again. It takes three hours to put it in and fit it up properly; it then had to come out again. We found that some of the hard pipes that go from the oil cooler to the engine didn't fit properly; they never fitted properly, and that's why they'd been in poor condition, chattering away on bulkheads and things.

"So, the process of putting the aeroplane together was actually becoming a big job and quite tedious. Once all that was done, the number one priority was to get some engine runs out of it and get the systems all working; making sure that the coolant system was working properly, the propeller was working properly with all the controls in the cockpit, that the oil system was okay, and we got a clean engine run out of it. We wanted to sling the fuel tanks in, put some fuel in it and get some decent engine runs. You can never really tell what these things are like until you get them on the aeroplane. Once the engine installation was all working, we moved on to seeing if it was taxiing okay, the pneumatic system all worked, and then we started on the completion of the aeroplane and the final inspections. We got a test permit on it once we had done all our paperwork, and got the CAA to issue a flight test permit to do some general handling tests on it. I normally hope to do about ten hours with an aeroplane like this, to shake it down, to start looking at all the new bits and pieces, how everything is working . . . a lot of removing of panels and seeing that everything is not chattering itself away with some unusual wear going on in places. You do some mini calculations, you see what the rate of climb is, that the fuel pressure and oil pressure and all the temperatures are all balancing up nicely, and that there is nothing that's gonna go amiss. You settle down nicely and not give it too

much of a hard time first off. You do need to give it a bit of a fistful on takeoff. There's no point hanging around near the ground. You do some cycling checks of flaps and undercarriage and make sure that's all gonna work properly and that there are no particular leaks anywhere . . . a good shakedown and good monitoring when it comes back. Finally, one gets it to a point at which everything is working and nothing is going amiss. You then fill the flight test document in for the CAA and they sign the aeroplane off.

"A restoration like that is pretty extensive, a lot of things in balance there that you've got to take on pretty carefully, not least the final inspection which has got to be in-depth and everybody has to be happy with it. That is the key problem, that inspection and making sure everything has gone back as it should. There's been some huge gaps between doing things. People forget what they've done. It's documented, but the guys who are constantly working on it don't get the opportunity to tick every box all the time. They have to use good engineering principles."

Jonathon Whaley made the first post-restoration flight of AR213 on Monday, November 12th 2007. Tony Bianchi: "The last time I flew AR213 was the 21st of April 2001, and prior to that, neither Jonathon or I had done much Spitfire flying in the preceding five years, so we were both a bit out of practice. But I have done at least 300 hours flying on that aeroplane over the years, and Jonathon has got quite a lot of time on it as well. He commented [after the first post-restoration flight] that there were

some small differences, but not dramatic ones. I found that there were some differences too. It's got more torque on take-off; definitely a lot more torque reaction than [it had] with the Dowty Rotol four-blade propeller. Possibly, the fact that the engine is overhauled and has a bit more power helps as well. In flight, it gives the effect of being a lot more solid and more taut. Before, it was all very loose. Everything worked quite nicely; everything was sort of light and agreeable. This is slightly different, almost like you stand back from it and think that, actually, there are some changes here that I'm not quite sure about. Its normal flying and handling characteristics are pretty much the same, but it feels slightly heavier. Still, when you don't fly an aeroplane for a long time, and you fly a lot of light aircraft, and your own strength is different anyway over that period of time, it actually feels slightly heavier, but then that settles down. It's a bit more pitch-sensitive, possibly because we've now got a heavier propeller. There are some aerodynamic changes too. The exhaust system, rather than having the twelve stubs on it, now has aerodynamically-shaped exhaust stacks on it, so there may be some airflow differences there. We've also got to counteract the increased forward centre of gravity with the heavier propeller. We've now got a heavier radio pack and different weights in the back; the centre of gravity is pretty much the same as it was, but the aeroplane is heavier so there is more momentum.

"We have some slightly rough running at the moment, which we think is propeller related. The engine seems fine, so once that's sorted we can play around with the power a little bit more. It's not really rough-running; actually it's more of a very low mode of vibration which we think is balance in the propeller or the spinner.

"Unlike with the original Mark IX / Mark V-type windscreen and hood, you're immediately in a different aeroplane; you have a much more sloping glass that you are looking through. More light seems to be coming in because it's taller, but the quarter light area, which is moulded on this particular screen, has very poor optics so it's distorted as you look lower into the corner of the screen. Unfortunately, that's the area you look at when you're on the ground; it's the only area you've got to look at. Alternatively, with the bubble-sided hood, you could put your head into the bubble area, so it made taxiing, or landing, or anything with the hood shut, slightly easier because you could actually angle around with your head inside. Now, you can't do that. You haven't really got the room. In the low-level and ground area, you are aware that it's narrower. You can't move your head about and you can't see so well. That makes one slightly uneasy. I hardly ever opened the hood before. I just kept it shut the whole time. Sometimes, if it was a lovely summer evening, I would open the hood and cruise back from somewhere very slowly. For the moment it's fine and we're getting on with it all right. We'll get used to it.

"In the air, once you are in cruise flight, you're not bothered about the windscreen because the aeroplane flies so much more nose-down. You tend to look through the area that is not so optically deformed, and then the hood doesn't matter anyway because you're only turning your head slightly; you're not moving it from side to side very much. The controls are slightly more stiff than they used to be, but that's to be expected as we've got new bearings and cables and whatever. Its general handling, if anything, has slightly better performance, which is a surprise to us. It's cruising around faster than it used to at the same power settings, which is an interesting point.

"We haven't checked its rate of climb yet because we are being easy on the power. I suspect the rate of climb will probably be as good, but not better than it was with the four-blade propeller, but I may be wrong.

PILOT JONATHON WHALEY TAKING AR213 FOR THE FIRST FLIGHT AFTER THE RESTORATION.

It's generally faster; it doesn't slow down so easily. The four-blade propeller had more disc area, so when you pulled the power back it tended to slow it up quite a bit; in fine pitch it was a quite efficient propeller brake, whereas this [three-blade propeller] isn't quite so efficient. Right down to the blade root area it's almost round anyway, so it's probably not producing any lift or drag. It's the outer part of the propeller that does that, from midway out. So, it's more slippery, which makes you plan what you are doing a little better, especially on a small airfield. You need to be a bit more organised in getting the pattern speed off, getting the undercarriage down and turning finals after a not very long downwind, so you don't have to grind the aeroplane in straight and can't see out the sides. The traditional curved approach is the only way to do it.

And, after thirty years of flying the aeroplane in knots, we are now in miles-an-hour as it should be, so all in all, you are having to change your thought process a little bit. It has a fractionally higher stalling speed and it feels a little bit more heavy; the whole aeroplane has a little bit more energy in it; it's more solid and a little less like the Mk Ia it used to be. It's very lively, still a very light aeroplane. It's got lots of power so it gets going quickly. I always relate to a Mark IX or a Mark XVI, the next stage on, as being a bit of a lumbering aeroplane compared with the Mark I. If anything, this one is a little bit better now than it used to be. It feels like a slightly later aircraft; like you've left the late thirties behind and you've gone into the mid-forties. It's become a bit more of a muscle plane.

"So, there are some changes. The biggest adjustment has to do with the approach and set-up to land. It's perfectly all right; it's relatively docile in those areas, but it's different. And we've gone back to the old brake system, which is not very good. You have to be quite careful. A few big applications of the brake on landing and you have to be very careful taxiing in afterwards; there's not much [brake] left.

"As to the flight testing now, we have to do a full flight-test programme. We need to get the aeroplane up to its maximum permissible speed, Velocity Never Exceed, VNE. We need to do a max continuous-power rate of climb up to ten thousand feet. I think, in the ways that we've changed it, we need to do some comparison flights to see if it really matches up to a standard Mk Ia Spitfire with this propeller configuration. So we'll do some flat-out

LEFT: JONATHON WHALEY ON THE
DAY OF THE FIRST FLIGHT; BELOW:
AR213 AIRBORNE AGAIN; BELOW
LEFT: TONY BIANCHI; BELOW RIGHT:
PERSONAL PLANE SERVICES CHIEF
ENGINEER TOM WOODHOUSE.

PAINTING, DECORATING, AND ADDING THE CAREFULLY RESEARCHED MARKINGS TO THE FULLY RESTORED AR213 MK 1A SPITFIRE. NEXT SPREADS: THE FULLY-DRESSED FIGHTER IN THE AIR OVER OXFORDSHIRE.

runs with it to see what the book speed is. We need to get a little bit more into just doing some normal aerobatics with it; the sort of stuff that we are used to, to see if there are any changes there. My guess is there will be some changes in that, with the Rotol propeller, as the aeroplane slowed up and came over the top of a loop, you hardly used any rudder, as the torque effect started to take you off. With this three-blade propeller, my guess is there will be a little more torque at the top of the loop, so you may have to use a little bit of rudder up there at the slow part of the loop. It will be interesting to see if there are any changes like that.

"We'll just do a normal flight-test programme, do some overshoots with it, the rate of climb check, a fair amount of stalling in clean and dirty configuration. Before we do any of that, we want to settle all of our adjustment snags to make sure the aeroplane actually flies hands-off, any vibrations are gone, all the bits and pieces that we need to make sure actually work properly to achieve the best performance . . . are all settled.

"How does it feel, the first time you take it back into the air again after such a restoration? I've done a few before. Basically, you've just got to shut off to everything around you. Having flown this aeroplane a lot, I just click back into the way it used to fly; I just fly it as it was. Oh yeah, you're always slightly pensive that something is gonna go amiss. When you're dealing with early stuff like that, you can never be entirely sure, and all the things that have been disturbed . . . that haven't been disturbed for years, they may suddenly start giving a problem, so

you've got to be on the ball. The problem with taxi runs is you run out of airfield. Once you've opened up the throttle you might as well keep going. When you're finishing off, you make sure the aeroplane is finished and done and everything works and all your brake pressures are correct and the undercarriage has been up and down properly and all the adjustments have been properly done. Then you go for getting airborne with it.

"It's nice to be flying the Spitfire again. It's nice to be flying a proper Spitfire again. It's quite a friendly old thing in a way. When you don't fly fighters a lot [in recent years], you have to take stock and say, 'I'm still in a lot of energy here, a thing that's exceeding the performance of all the other aeroplanes that I'm going to find around in the same part of the sky, so I've got to have my wits about me and I have to realise that I need a lot more room than what I'm normally used to. There is so much energy in these aeroplanes. They bash through any weather conditions quite quickly. You travel a long way very, very quickly, so you've got to get your head around that sort of thing. I don't find it too much of a problem because I did enough of it in the past, but I wouldn't particularly want to fly out of here [Booker] with low visibility and a little bit of rain, with the grease from the propeller on the windscreen and all the sort of little issues, and then have a problem as well. You know that it's waiting there to haunt you. That's why you fly on nice days with the wind down the runway and it's all nice and friendly."

ACKNOWLEDGMENTS

The author is grateful to the following people for their kind help in the development of this book and/or the use of their quoted and other material: Willis M. Allen, Jr., Sophie Armfield, Peter Arnold, David Atcherley, Steve Atkin, Peter Ayerst, Lord Balfour, Geoff Barlow, Antony Bartley, Malcolm Bates, Raymond Baxter, Julien Bersheim, George Beurling, Pia Bianchi, Tony Bianchi, Tony Bird, Austin John Brown, John Burgess, Allan Burney, Winston S. Churchill, Pierre Clostermann, Rachel Connolly, Neil Cox, David Crook, Budd Davisson (Airbum.com), Paul Day, Alan C. Deere, John Dibbs, Bob Doe, Hugh Dowding, Hugh Dundas, Steven Faulkner, Flight Journal, Stephen Fox, Isabella Freire, Luis Freire, Adolf Galland, Richard Gauntlett, Carolyn Grace, Nick Grace, Stephen Grey, Peter Haining, Alex Henshaw, Richard Hillary, Steve Hinton, Eric Holloway, Willem Honders, Gareth Horne, Margaret Horton, Pat Horton, Fanny Lucy Houston, P. H. Hugo, P. G. Jameson, H. E. Jones, José Jorge, Hargita Kaplan, Neal B. Kaplan, Brian Kingcome, Leslie Kingcome, James H. Lacey, Ludovic Lindsay, Ian Lloyd, Ian Lloyd, Adolph Malan, Peter March, Dennis Marsden, Eric Marsden, Margaret Mayhew, Mass Observation, Judy McCutcheon, Rick McCutcheon, Charles McMaster, George McMaster, Frank McNamara, Chris Meecham, Dr. Gordon Mitchell, R. J. Mitchell, Alan Moorehead, John Myers, Zdenek Ondracek, Alan Geoffrey Page, Sels Paul, Bob Pond, Bob Poynton, Jeffrey Quill, Günther Rall, Roel Reijne, Derek Robinson, Luis Rosa, J .S. Scott, William Shakespeare, W. M. Skinner, Richard Squibb, Franco Tambascia, George Thomas, Sam Thomas, Peter Townsend, George Unwin, Vickers-Armstrong, John Ward, Denis Le P. Webb, Jonathan Whaley, Allen Wheeler, Ray Wild, Willie Wilson, Tom Woodhouse, Tracey Woods.

BIBLIOGRAPHY

Andrews, C. F. and Morgan, E. B., Supermarine Aircraft Since 1914, Putnam

Barnato-Walker, Diana, Spreading My Wings, Grub Street, 2003

Bekker, Cajus, The Luftwaffe War Diaries, Doubleday & Co., 1968

Beurling, George & Roberts, Leslie, Malta Spitfire, Greenhill Books, 2002

Bickers, Richard Townshend, Ginger Lacey Fighter Pilot, Robert Hale, 1962

Bishop, Edward, The Battle of Britain, Allen and Unwin, 1960

Collier, Basil, Leader of the Few, Jarrolds, 1957

Collier, Richard, Eagle Day, Pan Books, 1968

Crook, D. M. Spitfire Pilot, Faber & Faber

Deere, Alan C., Nine Lives, Coronet Books, 1959

Deighton, Len, Fighter, Ballantine Books, 1977

Dibbs, John and Holmes, Tony, Spitfire Flying Legend, Osprey Aviation, 1996

Doe, Bob, Bob Doe Fighter Pilot, CCB Aviation Books, 2004

Dundas, Hugh, Flying Start, St Martin's Press, 1989

Franks, Norman, Sky Tiger, Crecy Books, 1980

Freeman, Roger A., Mighty Eighth War Diary, Jane's, 1981

Galland, Adolf, The First and the Last, Ballantine Books, 1954

Gallico, Paul, The Hurricane Story, Four Square Books, 1967

Gelb, Norman, Scramble, Michael Joseph, 1986

Glancy, Jonathan, Spitfire The Biography, Atlantic Books, 2006

Haining, Peter, The Spitfire Log, Souvenir Press, 1985

Hall, Roger, Clouds of Fear, Coronet Books, 1975

Henshaw, Alex, Sigh For A Merlin, Hamlyn, 1979

Hillary, Richard, The Last Enemy, Pan Books, 1942

Jackson, Robert, Spitfire, Parragon, 2005

Johnson, J. E., Wing Leader, Chatto & Windus, 1956

Kaplan, Philip, Fighter Pilot, Aurum Press, 1999

Kaplan, Philip and Collier, Richard, The Few, Blandford Press, 1989

Kaplan, Philip and Saunders, Andy, Little Friends, Random House, 1991

Kingcome, Brian, A Willingness To Die, Tempus, 1999

Lloyd, Ian, Rolls-Royce The Merlin At War, Macmillan Press, 1978

Lyall, Gavin, The War in the Air, Ballantine Books, 1968

Mason, F. K., Battle Over Britain, McWhirter Twins, 1969

Middleton, Drew, The Sky Suspended, Longmans, Green and Co., 1960

Mitchell, Dr Gordon, Schooldays to Spitfire, Tempus, 1986

Orde, Cuthbert, Pilots of Fighter Command, George G. Harrap, 1942

Oxspring, Bobby, Spitfire Command, William Kimber, 1984

Page, Geoffrey, Tale of a Guinea Pig, Pelham Books Ltd., 1981

Priestley, J. B., Britain Speaks, Harper & Brothers, 1940

Quill, Jeffrey, Birth of a Legend The Spitfire, Quiller Press, 1986

Robertson, Bruce, Spitfire—The Story of a Famous Fighter, Aero Publishers, 1960

Robinson, Derek, Piece of Cake, Alfred A. Knopf, 1983

Rudhall, Robert, Battle of Britain The Movie, Victory Books, 2000

Smith, Richard, Al Deere, Grub Street,

2003

Tidy, Douglas, *I Fear No Man*, Macdonald & Co., 1972

Townsend, Peter, *Duel of Eagles*, Simon & Schuster, 1970

Townsend, Peter, *The Odds Against Us*, William Morrow, 1987

Walker, Oliver, *Sailor Malan*, Cassell & Co., 1953

Webb, Denis Le P., *Never A Dull Moment*, J & KH Publishing, 2001

Wheeler, Allen, *Flying Between the Wars*, GT Foulis & Co Ltd, 1972

Willis, John, *Churchill's Few*, Michael Joseph, 1985

PICTURE CREDITS

Photographs by the author are credited PK; photos from the author's collection are credited AC; the jacket front panel is by Allan Burney; jacket back panel photos, from top left to bottom are credited to: Allan Burney, Vickers-Armstrong, Allan Burney, AC, and Julien Bersheim. jacket front flap: Luis Rosa, jacket back flap: Allan Burney, P3: PK, PP4-5: Austin John Brown, PP6-7: Julien Bersheim, PP8-9: Stephen Fox, PP12-13: John Myers, P14: AC, P16: PK, P18 all: AC, P19 left: A. Deere, centre: PK, right: PK, P20: Vickers-Armstrong, PP22-23: Vickers-Armstrong, P24: AC, P26: Vickers-Armstrong, P28: Cuthbert Orde, P29: AC, PP30-31: Willem Honders, P32: AC, P34: courtesy Tony Bianchi, P35: courtesy Tony Bianchi, PP36-37: courtesy Tony Bianchi, PP38-39: AC, P41: AC, P42: AC, P43: AC, P44: AC, PP46-47: Frank Wootton, P48: AC, P50: AC, P51 both: AC, P52: AC, P55: AC, P56 all: AC, P57 top: AC, bottom: PK, PP60-61: Zdenek Ondracek, PP62-63: Allan Burney, P65: Vickers-Armstrong, P66: Vickers-Armstrong, P69: Allan Burney, PP70-71: Allan Burney, PP72-73: Allan Burney, PP76-77: Allan Burney, PP78-79: AC, P80 left: Bundesarchiv, centre: AC, right: AC, P82: AC, P83 top both: AC, bottom: courtesy Andy Saunders, P86 both: AC, P87 all: AC, PP88-89: Michael O'Leary, PP90-91 all: Cuthbert Orde, PP92-93: PK, P94: AC, P95: AC, P97 all: AC, P98: National Portrait Gallery-London, P99 all: AC, PP100-101: Luis Rosa, P102: courtesy Tony Bianchi, P104 top: PK, bottom both: AC, P106: Bundesarchiv, PP108-109: Allan Burney, P110: AC, P111: PK, P112: AC, P113: courtesy Eric Marsden, P114 top: PK, bottom: AC, P116: AC, P119: AC, PP120-121: Roel Reijne, PP122-123: AC, P124 all: AC, P126 all: AC, PP128-129: AC, P131: courtesy Oscar Boesch, P132: AC, P135 all: AC, P137: AC, P138: AC, P139: AC, P140: AC, P141 both: AC, PP142-143: AC, PP144-145: courtesy Merle Olmsted, PP146-147: AC, P149: courtesy of Alex Henshaw, P151: Vickers-Armstrong, PP152-153: Vickers-Armstrong, P154: AC, PP156-157: Imperial War Museum, PP158-159: Stephen Fox, PP160-161: PK, P162: PK, P163: PK, PP164-165: PK, P166: AC, P167: AC, P168: PK, P169: AC, PP170-171: AC, P172: PK, P173: AC, P174: AC, P176: QC, P177: AC, P178: AC, P179: AC, P180: AC, P181: AC, P182: PK, P183 both: AC, PP184-185: Gareth Horne, P186 both: PK, PP188-189: Allan Burney, P190 both: PK, P191 both: PK, P192 both: PK, PP194-195: PK, PP196-197: PK, P198: PK, PP200-201: PK, PP202-203: PK, PP204-205: Allan Burney, P206 all: Tom Woodhouse, P209: AC, P211 both: PK, P212: PK, P213: PK, P214: PK, P217: PK, P218: PK, P219 all: PK, PP220-221 all: PK, PP222-225, Allan Burney, PP226-227: Austin John Brown, P229: PK.

A STAINED GLASS WINDOW GIVEN BY NO. 11 GROUP IN TRIBUTE TO THE CONTROLLERS WHO HELPED GUIDE THE SPITFIRE AND HURRICANE PILOTS TO INTERCEPT THE GERMAN RAIDERS ON BRITAIN IN THE SECOND WORLD WAR.